WHEN FRACKING COMES TO TOWN

WHEN FRACKING COMES TO TOWN

Governance, Planning, and Economic Impacts of the US Shale Boom

Edited by Sabina E. Deitrick and Ilia Murtazashvili

CORNELL UNIVERSITY PRESS   ITHACA AND LONDON

First published 2021 by Cornell University Press

Library of Congress Cataloging-in-Publication Data

Names: Deitrick, Sabina E. 1957– editor. | Murtazashvili, Ilia, 1975– editor.
Title: When fracking comes to town : governance, planning, and economic impacts of the US shale boom / edited by Sabina E. Deitrick and Ilia Murtazashvili.
Description: Ithaca [New York] : Cornell University Press, 2021. | Includes bibliographical references and index.
Identifiers: LCCN 2021009261 (print) | LCCN 2021009262 (ebook) | ISBN 9781501760983 (hardcover) | ISBN 9781501760990 (paperback) | ISBN 9781501761003 (ebook) | ISBN 9781501761010 (pdf)
Subjects: LCSH: Hydraulic fracturing—Economic aspects—United States. | Hydraulic fracturing—Government policy—United States. | Hydraulic fracturing—Law and legislation—United States. | Hydraulic fracturing—Social aspects—United States. | Shale gas industry—Economic aspects—United States. | Shale gas industry—Government policy—United States. | Shale gas industry—Law and legislation—United States. | Shale gas industry—Social aspects—United States.
Classification: LCC HD9581.2.S53 W44 2021 (print) | LCC HD9581.2.S53 (ebook) | DDC 338.2/72850973—dc23
LC record available at https://lccn.loc.gov/2021009261
LC ebook record available at https://lccn.loc.gov/2021009262

Contents

Illustrations

Preface

For decades, geologists knew that shale contains massive amounts of natural gas. It was not worth much economically until drillers in Texas figured out a technique that came to be called "fracking." This new technique enabled drillers to mine shale gas profitably. The result has been a shale boom that transformed the energy landscape in the US, but also the world. For some, this meant jobs and new economic opportunities, but for others, the shale boom has been accompanied by controversy, such as concern about the economic, environmental, and health impacts of shale gas development, as well as whether citizens have been able to exercise their voice during the shale boom. This volume brings together scholars to consider the governance, planning, and economic impacts of the US shale revolution. It offers an explicitly interdisciplinary perspective grounded in the local experience with shale gas development. Its chapters provide a granular account of the extent to which there has been effective governance of the shale boom, how planners in boomtowns have responded to the rapid rise in shale gas development, and what have been the economic consequences of the shale revolution in different places.

Acknowledgments

This volume began with a workshop held in 2015 seeking to understand the rapid rise in US shale gas production from an interdisciplinary perspective. Its motivation was that understanding something as complicated as the rapid rise in shale gas production required moving beyond disciplinary confines to consider the governance (including legal), planning, and economic consequences of shale gas. After we gathered, many of the participants agreed to write chapters on the themes addressed in the workshop. We also solicited chapters from experts on related subjects of importance. Over the next years, as fracking continued, the authors gained additional perspective on its consequences, as well as on local response. This volume is a result of these efforts to offer a thorough, balanced assessment of the shale gas revolution from a variety of perspectives.

We thank the Graduate School of Public and International Affairs (GSPIA) at the University of Pittsburgh for support, especially Dean John T. S. Keeler. We have also benefited from the support of the Shale Gas Governance Center at GSPIA, especially from the insights of Jeremy G. Weber and our excellent students who have worked in the center, Max Harleman, Insik Bang, and Ion Simonides. We also would like to thank the University of Pittsburgh Center for Social and Urban Research (UCSUR) and Eric Beckman and Gena Kovalcik of the Mascaro Center for Sustainable Innovation at the University of Pittsburgh for early support of this project. Anna Aivoliotis at UCSUR provided her expert editing skills to our volume. Boris Michev, the liaison librarian of GIS, Urban Studies, Social Sciences, and Maps at the University of Pittsburgh library, did the greatly appreciated work of generating the publishable figures for this volume. We especially thank the editor we initially worked with at Cornell University Press, Michael McGandy, for supporting the project, and with Jim Lance for continuing it. Clare Jones provided much-needed editorial support. We also thank four anonymous reviewers for excellent comments at early and later stages that improved the project a great deal.

Research for chapter 10 was developed with support from the US Small Business Administration (Contract #SBAHQ-13-C-0016). Local support was provided by the Northwest Pennsylvania Regional Planning and Development Commission and the Northwest Pennsylvania Partnership for Regional Economic Performance. Research support was provided by Brendan Buff, Sarah Gutschow, and Wen Sun of the Center for Regional Competitiveness.

WHEN FRACKING COMES TO TOWN

BEYOND THE BOOM

Sabina E. Deitrick and Ilia Murtazashvili

By now, just about everyone has heard about "fracking," the extraction of natural gas through the process of hydraulic fracturing. Those who hoped for greater federal involvement in fracking were disappointed by President Barack Obama, who did little to stop the shale boom. President Donald Trump promoted fracking and the oil and gas industry. There is also an apparent ideological divide with fracking: proenvironment Democrats, women, and people living in urban areas are more likely to oppose fracking (Davis and Fisk 2014), while many in the business community, including many state and local government officials, support fracking. Though one of President Joe Biden's early executive orders imposed a moratorium on fracking on federal lands, he has not indicated opposition to fracking elsewhere. Fracking is unlikely to go away, leaving states and municipalities to deal with its consequences.

This book considers the boom and bust aspects of the recent shale boom. It acknowledges that in the long run, most communities end up with extensive costs from extraction, especially on the environmental and public health fronts. For example, researchers who conducted a cost-benefit analysis in the Pennsylvania, Ohio, and West Virginia basin found that shale gas has massive costs in human health and environmental damage—much more than the economic benefits (Mayfield et al. 2019a). Their conclusions support much work in public health, but unlike most studies that focus on jobs and royalties—the most important benefits from fracking—the study mentioned above considers the costs from a longer-term perspective comparing the effects of degraded air on health and longevity against sustainable job creation. The results are grim: the longer the time horizon, the more evident the health costs. Spatial inequities, especially racial and income inequities, are another challenge, as downwind forces of air pollution beyond the local extracting area are more likely to affect groups of lower socioeconomic status (Mayfield et al. 2019b).

As the studies just referenced indicate, the debate over fracking is about uneven and unsustainable economic growth. Herman Daly (2005) argues that one of the problems with today's economics is that it assumes the global economy exists in a void, rather than in a biosphere that supports it. Daly argues that this thinking contributes to *un*economic growth, which produces "bads" faster than goods, making us poorer, not richer. For many, fracking is uneconomic growth that eats up the ecosystem and human health at the expense of tax revenue that is used for investment in infrastructure.

One important related challenge is how to conceptualize costs. For climate change, as well as human health, these are catastrophic, low probability possibilities with massive costs (Weitzman 2009, 2011). Analyzing those costs is challenging in the usual economic approach to costs and benefits, which seeks to place a dollar value on all relevant impacts of an economic activity.

One option is to keep the shale gas in the ground. A team of University of Pittsburgh public health researchers has done much work to show why it makes sense to leave it all in the ground (Goldstein, Bjerke, and Kriesky 2013), as well as the importance of bringing more voices into the discussion of shale gas development, including public health researchers (Goldstein 2014; Goldstein, Kriesky, and Pavliakova 2012). New York State's 2014 ban on hydraulic fracturing, which was made permanent in 2020, was one important policy on this front.

Of course, the shale train is already out of the station. Regardless of where one stands on the issues raised above, for many in local communities, it is necessary to come up with a policy for shale gas development, as well as to mitigate its consequences. Much of the research noted above adopts a pragmatic approach to policy, considering what can be done using more traditional economic solutions. The Mayfield et al. (2019a; 2019b) research suggests a tax on production; Landrigan, Frumkin, and Lundberg (2020), reflecting on the health and environmental costs of plastics production—called "cracker" plants because of the technique used to create plastics during the fracking process—that often come with shale gas development, suggest eliminating subsidies for shale gas development. Each applies a usual economic solution to a negative "externality," or social cost, which is to tax it or remove a subsidy.

Even if one accepts several premises critical of fracking—that the bust is inevitable, that all resource booms are not sustainable, and that in the end, local economies will be left mainly with health and environmental costs—it is necessary to consider the response to fracking. In this regard, Daniel Raimi's (2017) *The Fracking Debate* offers an insightful perspective on how to approach these issues. Raimi's discussion of the precautionary principle—the idea that we ought to wait to develop things until we know a lot about the consequences—is especially apt: "If the precautionary principle is the lodestar by which we navigate risks and rewards, it quickly becomes difficult to justify using any number of new technologies" (7). The solution, he argues, is to have a well-informed public debate on fracking going forward.

Our volume takes up where Raimi leaves off. We want to understand how communities have dealt with the shale boom, including its economic costs. Our interest lies in how planners and government officials have responded to the challenges and opportunities from natural resource extraction. Many of these challenges may be exacerbated as state and local governments deal with the financial fallout from the COVID-19 pandemic, as well as the potential for increased pressure for shale gas development as a way to boost the economy and keep prices low as the curve flattens. It is hoped that we can provide insight into the specific legal, governance, planning, and economic challenges faced, as well as an understanding of how communities have responded when they often had no choice in the matter.

How Did We Get Here?

The shale boom refers to the dramatic rise in US shale gas production starting in the late 1990s and continuing into the 2020s. Most of this natural gas comes from shale, which is a type of semipermeable rock. Shale formations contain vast quantities of natural gas, but typically lie thousands of feet below the earth's surface. The geographic scale of these shale plays is extensive, as figure I.1 shows.

For many decades, the natural gas housed in shale was not worth much economically. Its geological features prevented the use of conventional techniques of drilling downward to profitably mine it. The economic profitability of shale gas changed as a result of the entrepreneurial activities of drillers working the major shale basin in Texas, the Barnett Shale. These drillers refined a technique to extract the gas trapped in shales through a process called high-volume hydraulic fracturing. This technique, which is popularly known as fracking, works by injecting water at high velocities to fracture shale deposits, which then allows the gas to flow to the surface where it can be profitably mined. The combination of hydraulic fracturing, directional drilling, and the right mix of chemicals had the effect of a major technological innovation on natural gas extraction that enabled the shale boom (Fitzgerald 2013).

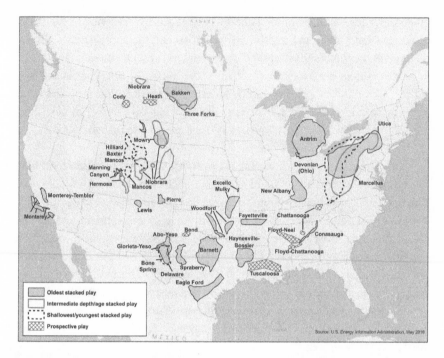

FIGURE I.1.   United States shale plays. United States Energy Information Administration, May 2016.

The innovation in techniques used to mine shale gas was an interesting beginning to a still-unfolding story. For some, fracking represented new economic opportunities. Many folks in communities that lived through the ups and downs of conventional oil and gas production found hope and maybe dreams in shale gas. For them, it was an economic blessing. For others, shale gas was a Faustian bargain. In 2012, Hollywood made a movie about the shale boom, *Promised Land*, which depicted landmen, whose job was to get the owners of mineral rights to sign leases with the gas companies, preying on the hopes and dreams of would-be lessees. The movie viewed the landmen as dishonest, deceiving owners into signing away their mineral rights for less than they were worth and without rightsholders having a clear understanding of the effects of fracking on their environment and health. Journalist Tom Wilber's (2012) book, *Under the Surface*, captured the concerns of community members in the Marcellus Shale region through a remarkable series of interviews with folks who were touched by the shale boom. These accounts depicted the landmen, and the shale boom more generally, in an unfavorable light. A documentary film, *Gasland*, focused on residents' experiences with drilling and stoked fears of methane in groundwater.

Beyond consternation surrounding landmen, many raised other important economic concerns, such as what fracking might do to property values, with fear of groundwater contamination devaluing home values, as well as concern about boom-bust cycles, in particular what might happen once shale production subsides. Some early studies did not find a strong link between fracking and methane in groundwater (Vidic et al. 2013), but since then, a vast literature has considered environmental and public health concerns centered on air and water contamination. Concern has expanded to the harms from air pollution near wells, such as evidence of lower birth weight babies for women living in proximity (within two km) to shale wells during pregnancy (Currie, Greenstone, and Meckel 2017). Seismic effects related to underground injection wells and wastewater disposal are another concern (Hand 2014; Davis and Fisk 2017), especially in Oklahoma, where there have been hundreds of small earthquakes near drilling sites, and even in the Marcellus and Utica Shales, where the first small quakes in Pennsylvania were recorded in 2016 (Frazier 2017). Other impacts affect quality of life issues in communities affected by a resource boom, including increases in crime requiring increases in police services (James and Smith 2017). There is also concern about damage to local roads from truck traffic generated from shale gas drilling, as well as "fraccidents"—the increase in accidents because of more trucks on the road (Xu and Xu 2020).

Economic impacts for local governments and residents are another important story. Some work finds that homes that depend on well water near well pads lose property value (Muehlenbachs, Spiller, and Timmins 2015). Black, McCoy, and Weber (2018) consider the consequences of impact fees, which are fees assessed on shale producers that serve a similar purpose as a tax, but are tailored to address externalities associated with fracking to pay environmental and infrastructure costs. Their results, which used a natural experiment to compare Pennsylvania to similar geographic areas in Ohio and West Virginia, found that the impact fee reduced leasing activities, which they found to be counterproductive given that the state was experiencing an economic downturn at the time. The shale boom has also led to a rich literature considering the diversity of responses at the state level to challenges posed by shale gas development (Majumdar 2019; Fisk 2016; Davis 2014). Perhaps unsurprisingly in light of the complexity of issues surrounding fracking, public opinion is often sharply divided, as well as dependent to an extent on the information available to individuals (Arnold, Farrer, and Holahan 2018). Relatedly, many questions have been raised about state disclosure practices regarding the shale industry (Fisk and Good 2019; Fisk 2013).

For critics and skeptics, shale gas production is a curse, not a blessing. However, the reality of the situation is that the shale boom is more complicated than these accounts suggest. The analysis of the shale boom requires consideration of the benefits and costs of hydraulic fracturing along many dimensions (Murtaza-shvili and Piano 2018, 2019b). The benefits include the potential consequences for economic growth, reducing unemployment, and increasing household income, most often the key drivers of public officials' support—and subsidies—for the industry. The costs include negative externalities, including health, environment, and infrastructure costs discussed above. The shale boom also raises many questions about the political response to fracking.

These concerns can be organized into three distinct yet interrelated categories: governance, planning, and economic impacts of the shale boom. Governance of the shale boom refers to the economic and political context within which the shale boom has occurred. Planning with respect to the shale boom considers how municipal officials have responded to the potential for a bust. The economic impacts of shale gas development refer to its impact on human well-being, including the extent of economic externalities. We introduce each of these perspectives below.

Governance of the Shale Boom

Part I of the book considers governance of the shale boom. Governance refers to the study of good order and working relationships (Williamson 2005). The quality of governance of the shale boom is a major theme in debates over fracking. Some question whether there has been much governance at all, likening the shale boom to a modern-day Wild West. Another question is whether the regulations for conventional oil and gas are even appropriate for the shale boom, which involves different regulatory challenges, such as more of a concern with groundwater and air pollution (Holahan and Arnold 2013). Other issues include the divergence of how states have responded, such as a massive increase in shale production in Pennsylvania and a ban next door in New York. The divergence of regulation in states sharing a shale play (see figure I.2) has led some to question the appropriateness of federal regulation (Arnold and Holahan 2014; Arnold and Neupane 2017). In the opinions of some, the political process is biased toward gas companies and other voices are excluded, resulting in protest (Gullion 2015). However, there is also a literature which suggests that the United States has quality governance institutions, including clear property rights and state and local governments with substantial regulatory capacity, that address many of these challenges (Harleman and Weber 2017; Murtazashvili 2017).

Part I of this volume addresses these questions of governance of the shale boom. Together, the authors of these chapters clarify the political, institutional,

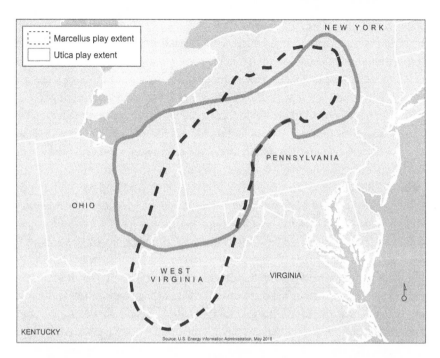

FIGURE I.2. Marcellus and Utica Shale plays. United States Energy Information Administration, May 2018.

and regulatory context within which the shale boom has occurred. In chapter 1, Ilia Murtazashvili and Ennio Piano consider popular accounts that the shale boom has occurred in an institutional and regulatory Wild West, with emphasis on how well the gold rush of the nineteenth century provides a useful metaphor to consider governance of fracking. Their argument is that both the gold rush that commenced in California in 1848–1849 and the shale boom are more orderly than is often presumed in the popular press and even in the academic literature. They argue that the United States benefits from a long history of private property rights and from market intermediaries, especially landmen, who enabled a rapid response to new economic opportunities. They also contend that the states are formidable regulatory bodies that have much experience with energy and so the diversity of regulatory responses can even be thought of as a virtue of the US federal system. More generally, they show that the notion of a fracking Wild West is inadequate when considering the robust, albeit imperfect, governance response to the shale boom.

Heidi Gorovitz Robertson, in chapter 2, examines how local governments in the three main states of the Marcellus Shale region, Pennsylvania, Ohio, and New York, have grappled with shale production. The reason to focus on these three states is that they share this massive shale play and their response to opportunities and governance challenges presented by fracking have varied tremendously. Robertson shows that local governments have attempted to control shale gas drilling and related factors in their municipalities through zoning and land use powers within the larger structure of governance, with states attempting to legislate complete control to create uniform statewide practice and regulation for companies in the industry, despite home rule provisions in all three states. State law in Pennsylvania and Ohio both attempted to preempt local ordinances. The analysis shows that decision-making varies by state constitution, statutes, and judicial rulings across all states, suggesting that the final land use and zoning roles for local governments are largely a factor of their state legal environments and what local level authority can be supported or not.

While Murtazashvili and Piano's chapter is optimistic about the prospects for shale governance, several contributions question the governance of the shale boom, including from a distributional perspective. In chapter 3, Ann M. Eisenberg tackles an often-ignored component of the shale gas revolution and fracking debate—that the Marcellus and Utica Shale plays are part of Appalachia. Her analysis of the law in Pennsylvania, Ohio, and West Virginia adds to the analysis in chapter 2 with its focus on landowners' and legal regimes' participation in environmental justice. She tracks the long history of environmental justice issues from Appalachia's coal dependence and the environmental and health costs the region paid as coal communities through the continued absence of robust environmental regulations in the fracking boom in often these same communities.

The fourth chapter in the collection, by Pamela A. Mischen and Joseph T. Palka Jr., takes a social constructivist approach to examine how local municipal officials in Pennsylvania have found ways to address the continued conflicts between environmental and health disadvantages and economic benefits. In analyzing the actions of local officials, they build on our understanding of formal and informal means of institutionalization of norms and knowledge and how divergent views of hydraulic fracturing are understood by local municipal officials in areas where drilling occurs. Their analysis focuses on narratives of local officials and which narratives are common versus which were divergent. Their results show that municipal officials confirm their ability to address issues compounded by fracking and that administrative responses were in place to confront the challenges that occurred. Their analysis also confirms other studies that show that in municipalities where drilling occurs, support and means to moderate conflicts are strong and work to reduce chances of negative zoning responses. Their work also examines differences between communities with zoning ordinances and those without zoning. Despite these differences, the authors found drilling in most communities has been relatively conflict free, with local officials playing a strong moderating role in the governance of shale gas drilling.

Planning for the Boom (and Inevitable Bust)

Part II of the volume focuses on the policy and planning challenges that the shale industry brought to state and municipal planning agencies. Planners have zoning and other powers to govern the process of economic development and natural resource extraction when fracking is exempt from federal regulations. Planners also face challenges with zoning and land development ordinances when confronting both resource boom and busts, and many areas are dominated by progrowth coalitions where developers work with local government to encourage prodevelopment policies (Molotch 1976), common in the fracking supply chain. However, there still remains substantial autonomy at the local level to regulate (Vogel and Swanson 1989). Planners as a result confront both opportunities and challenges to respond to the shale boom, such as how to account for the possibility of a shale bust and use revenues from the boom to address social costs, such as infrastructure deterioration.

A careful consideration of planning complements our analysis of governance and the economic impacts. Our analysis of governance focuses mainly on the structure of institutions—especially economic and legal rules—alongside the politics of fracking. Planning, in contrast, deals more with the working rules that govern the boom, as well as how to address potential consequences of a bust. In other words, planning is more on the policy and public administration side of the equation. In addition, while economic analysis often considers the impact of fracking, planners have to deal with those consequences at the local level and the impacts on residents, as well as deal with the "details" of the shale boom, such as how state revenues are spent among often dozens of competing priorities for such spending.

Our authors offer deep knowledge of the planning process through a set of detailed case studies of the planning response in different communities facing a prospective bust from shale gas extraction. At the same time, these chapters consider the big picture of planning, such as how this boom compares to other ones of longtime, continuing significance for planners and students of planning and the policy process.

In chapter 5, Adelyn Hall and Carla Chifos reflect on the resource curse and the boom-bust consequences familiar in the rural United States with a case study of Bradford County in Pennsylvania's northern tier, which lies above part of the Marcellus Shale. As largely agricultural Bradford County transforms rapidly with natural gas extraction, boomtown models are compared, with the Bradford County current shale gas boom following along lines of previous energy boom-towns. The new energy development in Bradford County and its communities follows models developed over decades by researchers and planners studying boom and bust trends and development; but in contrast to more typical scenar-ios, Bradford County responded with new focuses on human capital investments, new partnerships with higher education institutions, and other capacity-building initiatives for local officials and community stakeholders to shift the power bal-ance to community development rather than reactive responses.

In chapter 6, Teresa Córdova and her group of students travel to Williston, North Dakota, another boom ready to bust three decades after an earlier energy boom-bust cycle in the 1980s. The advent of fracking in the Bakken oil field brings shale oil production into this volume, with a longer lens from which to analyze the changes and learning that can occur in resource rich boom-bust communities over multiple boom-bust cycles. The case of Williston shows that communities can address many relevant issues of the boom-bust cycle through planning and policy choices. The case study examines the crucial role that local and state officials and civic leaders have in mitigating the strongest forces of boom and busts. Williston learned how to use local planning for power when the shale oil boom was turning to a bust by 2014. The actions of local and state officials enabled Williston to move through a downturn in the industry and not become a "bust" community. With sound planning and economic development practice, coupled with important changes in taxes and regulations, the area did not experience the "wild rides" of the resource bust, but learned how social cohe-sion could sustain the downside of an extractive economy.

Understanding planning and impacts on smaller communities continues in "Local Planning in Beaver County and the Shell Cracker Plant," echoing themes developed in the preceding chapters as state progrowth forces reduced planning and environmental policy power at the local level. Sabina E. Deitrick and Rebecca Matsco analyze the planning and governance responses in a small community that was selected by state officials to be the new location for a multi-billion-dollar cracker facility. With little prospect of stopping the project, local officials and the community sought to expand their capacity in planning and public administra-tion through several concerted efforts to create benefits from the project that went beyond the actual project. Local officials moved beyond reactive stances in their will to act, showing that capacity itself is a fluid concept in planning and governance. The conclusions reflect the arguments developed in Part I on gover-nance, with the planning aspects of conflict mediated at the local level.

In chapter 8, Anna C. Osland and Carolyn G. Loh focus on the conflict for planners balancing sustainability with economic benefits generated by fracking. They conducted a survey of municipal planners across three hundred jurisdictions in Pennsylvania, North Dakota, Louisiana, and Colorado to understand how local residents and businesses evaluate the economic benefits of the shale industry, the differences between them, and what growth management adjustments or tools have been adopted in the wake of the boom from extraction. The survey results showed that respondent planners reported economic benefits for both local businesses and residents, with the benefits to businesses outweighing those to residents. They also found that these gains were associated with reduced housing affordability in their communities. Nonetheless, in all states the economic effects of fracking were seen as generating no effect or improving local economic conditions. Through their survey, they conclude that local planners have not often faced or confronted the trade-off in boom-bust communities between economic development and environmental protection.

Economic Impacts of Shale Gas Development

There is a huge debate over the economic consequences of fracking. In a widely cited and thorough research paper, Hausman and Kellogg (2015) find that fracking generates billions of dollars annually, especially under scenarios when fracking does the most damage to coal production. Another team of economists, Feyrer, Mansur, and Sacerdote (2017), finds that there has been a reduction of unemployment during the Great Recession and that there was a substantial increase in royalty rates in the county and area surrounding the county with hydraulic fracturing. The shale boom has also resulted in billions in royalty payments to the owners of mineral rights (Brown, Fitzgerald, and Weber 2016). These studies, which are increasingly using massive amounts of data to draw their conclusions, suggest that there was a good reason for many to be optimistic that their household incomes would increase because of shale gas production. There are also important economic consequences for public finance, including funding for schools. In Texas, the shale boom appears to have improved housing values, which in turn increased revenue for public schools (Weber, Burnett, and Xiarchos 2014). Yet it is still unclear whether the taxation of shale is economically efficient. One reason is that there is tremendous variation in the taxes imposed (Weber, Wang, and Chomas 2016). Another possibility is a fracking resource curse, which is lower employment in shale producing areas because of shale production (Weber 2014). A challenge is that resource booms are very long-run phenomena, and only now are we getting a good idea whether the boom and bust cycles of the 1970s and 1980s of conventional oil and gas are having a beneficial long-run impact (Jacobsen and Parker 2016).

Part III complements existing studies by further evaluating the economic impact of fracking. These include efforts to better understand the economic impacts with new data, as well as offering new perspectives on the economic analysis of the shale boom. Together, our economic analysis provides additional insight into how we assess the impact of shale gas development on the economy. It also provides insight into what we can call a regulatory knowledge commons: a pool of ideas about how to regulate, or not regulate, the shale boom (Murtazashvili and Piano 2019a). One of the benefits of a polycentric system of governance, with multiple avenues of experimentation with fracking, is that there is more knowledge about what may work well in regulating shale gas development, what does not work so well, and the ways in which regulators have dealt with political challenges that arise as shale production begins to increase.

Frederick Tannery and Larry McCarthy's work, "Local Labor Markets and Shale Gas," begins Part III. For many in the Marcellus Shale region, the promise of jobs was one of the biggest selling points for allowing fracking to proceed. As Hall and Chifos describe in Bradford County, PA, in chapter 5, and Mischen and Palko relate through their survey results in chapter 4, proponents point to employment growth as the major impact of the industry. Beginning with data from before the boom and continuing through the downturn to 2015, Tannery and McCarthy show that the impacts of the shale gas industry have contributed to economic growth in rural counties in Pennsylvania, and that in rural areas with long periods of economic disinvestment and job losses, the benefits have accrued in long-standing areas of decline. The analysis compares outcomes in high gas producing counties to others with no gas development, and though it does not conduct a cost-benefit analysis, its jobs approach to shale gas drilling in rural regions shows significant gains.

Following is "Shale Energy and Regional Economic Development Impacts in Northwest Pennsylvania," by Erik R. Pages, Martin Romitti, and Mark C. White, which picks up where Tannery and McCarthy's focus on direct job impacts leaves off. Here, the authors build a case for understanding and analyzing the supply chain of production, and, particularly, manufacturing and service spinoffs and spillovers that can impact older industrial economies in the heartland. The authors document the value chain for northwestern Pennsylvania, a region to benefit from Marcellus and Utica Shale drilling. Their work also covers the beginning through the boom and bust of the sector, to address economic impacts over time, however brief the boom-bust cycle in the region has been. The authors break the process down into three major phases of development and production to determine differences in industry impacts by each phase. The authors relate the opportunities and constraints in supply chain development through local sourcing through all three phases. While potential development is robust, additional resources from federal manufacturing assistance programs to state and regional economic development and planning efforts can add more value to the region's potential value chain.

Nicholas G. McClure, Ion G. Simonides, and Jeremy G. Weber, in chapter 11, consider the important question of what happens to wells, and whether the economic incentives are appropriately designed for inducing socially efficient behavior. State governments require operators of oil and gas wells to reclaim well sites at the end of a well's economic life. To encourage and fund proper reclamation, states require operators to set aside money in the form of a bond, which is forfeited to the state if the operator abandons a well without reclaiming it to state standards. The state can then use the money to pay contractors for reclamation. For bonding requirements to encourage reclamation—or give the states resources to do it—they must be set near the cost of reclamation. Estimates of the cost of reclaiming shale gas wells suggests that current bonding requirements are too low, especially in Pennsylvania, which has had the greatest expansion of shale gas development in the world. They use data from seventy-eight reclamation contracts for 1,205 improperly decommissioned oil and gas wells in Pennsylvania from 2005 to 2016 to better understand reclamation costs. The data permit documenting average reclamation costs for conventional wells and, most importantly, why some wells are far more expensive to reclaim than others. McClure, Simonides, and Weber use the relationships between well characteristics and reclamation costs to provide a better estimate of the cost of reclaiming the typical shale gas well in Pennsylvania.

In chapter 12, Gavin Roberts and Sandeep Kumar Rangaraju consider the dynamics of the Wyoming gas boom. Natural gas development boomed in Wyoming between 2000 and 2011 leading to a near doubling of gas production in the state. New technologies combined with high gas prices expanded the array of economically recoverable gas reserves. This led to expansion in Wyoming's economy relative to the rest of the United States, and inflation in Wyoming's tax revenues related to severance taxes, which account for approximately 40 percent of total tax receipts. This tax windfall led to growth in Wyoming's state government. However, more recent lower gas prices and the limited availability of low-cost gas drilling prospects in Wyoming have led to contraction in drilling and decreasing tax receipts. This led to a crisis in Wyoming's higher education system, and financial trouble in other state institutions. The state government is currently attempting to cut spending, but path dependence in budgets associated with "use it or lose it" incentives is making the task more difficult. The economy continues to benefit from the recent drilling boom, as prices remain historically low in a state where residents rely heavily on gas for heat in the winter. The primary short-term threat to the economy in Wyoming is spillover from budget cuts. This chapter provides historical context of gas production, analyzes economic impacts, and analyzes how those impacts differed between the private and public sectors in the state of Wyoming. It concludes by clarifying which sorts of policies make the public sector less susceptible to resource-related, short-term boom and bust cycles, and protect the private sector from the associated spillover.

In chapter 12, "An Economic and Policy Analysis of Shale Gas Well Bonds," Max Harleman considers a fundamentally important way governments seek to address the economic impacts of fracking through bonding requirements. Most states impose bonding requirements to incentivize operators to plug wells and restore surrounding land and water when they stop producing oil or gas. Bonding requires operators to set aside funds before drilling a well, which are forfeited to the state if the well is abandoned. The relatively high reclamation costs and potential environmental risks from shale gas wells have led state officials to consider whether current bonds are large enough to cover cleanup. Others argue that operators would respond to increased bonds by drilling fewer wells. Less drilling could reduce economic benefits received by state residents, such as those from wages and mineral royalties. This chapter aims to answer the question: how much money should state officials require shale gas well operators to set aside in the form of bonds to achieve a socially desirable outcome? The analysis distinguishes the categories of costs and benefits that state officials should consider when evaluating bonding systems and presents a cost-benefit analysis of bond systems in Pennsylvania. The findings reveal that current bonds are too low and do not match actual reclamation costs. The chapter recommends that state officials set bonds equal to best estimates of reclamation costs, a practice that will prevent significant costs borne by residents and without the loss of many economic benefits.

The final chapter of the book pulls together the conclusions of these case studies. The shale boom has created many studies on environmental and health impacts, but far less on what happens in local communities in practical terms of governance, planning, and economic impacts. These cases demonstrate the role and importance of local and state officials, civic leaders, and resident stakeholders in challenging—and often untested—conditions. This chapter offers recommendations and the role of continuing research on understanding the practice and impacts of shale gas production in local and state planning and governance.

Situating This Volume

It is useful to reflect briefly on the scope of the volume. First, the contributions focus almost exclusively on shale gas, with one chapter that considers shale oil extraction in the Bakken play in northwestern North Dakota. Shale oil is an important part of the energy landscape and will continue to be in the future, especially in North America, which has tremendous quantities of shale oil. Indeed, drillers in Alberta, Canada, were the first to begin experimenting with hydraulic fracturing to extract shale oil in the 1950s. Shale oil has not yet had the same impact on the US economy as shale gas nor impacted as many regions. Nonetheless, our analytical focus, which emphasizes consideration of energy booms from a governance, planning, and economic perspective, can enhance understanding of related types of unconventional energy extraction, including shale oil.

Second, our focus is on shale gas extraction in the United States even though there has also been substantial debate about fracking outside the United States. In Canada, for example, Albertans have promoted fracking, while Quebec put a moratorium on fracking from 2011 to 2017. In Europe, Poland pushed fracking, while France quickly banned it entirely in 2011. Unfortunately for Poland's fracking proponents, Polish shales are not especially productive, and so the recoverable shale gas turned out to be quite marginal, and production has not increased much. Britain temporarily stopped fracking, shutting down the only active facility in England, while Scotland and Wales had long halts. China's fracking has proceeded apace, especially in the Sichuan region, while Argentina's Vaca Muerta (which translates as "Dead Cow") shale basin has seen substantial investment. One reason why fracking in China is interesting is because the country leads the world in carbon dioxide emissions, in large measure because of coal-fired power plants. Fracking is viewed as a cleaner alternative to coal, and hence a way to reduce emissions. Part of our reason for leaving out such analysis is pragmatic: there has been such a wide variety of responses, and such a diversity of ways to evaluate the consequences of fracking, that a focus exclusively on the US shale gas boom (and inevitable bust) is not only reasonable, but in some ways necessary to understand the consequences of such a complex phenomenon. Even so, another reason to focus on the United States is that despite promise of an explosion of shale gas production around the world, the United States still leads the way by a large margin, and so it is the epicenter of studies to address how communities respond to fracking, as well as to analyze its impacts.

Third, the studies in our volume complement narrower, disciplinary studies characteristic of much of the economic and public health research on fracking. Economic studies offer precise insight into the impact of proximity to shale wells, how much money is changing hands, and the efficiency of the contracting process. Such studies provide fewer insights into the governance of the shale boom, including its institutional foundations, or the ways that planners have attempted to deal with the economic consequences of the shale boom.

References

Arnold, Gwen, Benjamin Farrer, and Robert Holahan. 2018. "How Do Landowners Learn about High-Volume Hydraulic Fracturing? A Survey of Eastern Ohio Landowners in Active or Proposed Drilling Units." *Energy Policy* 114:455–64.

Arnold, Gwen, and Robert Holahan. 2014. "The Federalism of Fracking: How the Locus of Policy-Making Authority Affects Civic Engagement." *Publius* 44 (2): 344–68.

Arnold, Gwen, and Kaubin Wosti Neupane. 2017. "Determinants of Pro-Fracking Measure Adoption by New York Southern Tier Municipalities." *Review of Policy Research* 34 (2): 208–32.

Black, Katie Jo, Shawn J. McCoy, and Jeremy G. Weber. 2018. "When Externalities Are Taxed: The Effects and Incidence of Pennsylvania's Impact Fee on Shale Gas Wells." *Journal of the Association of Environmental and Resource Economists* 5 (1): 107–53.

Brown, Jason P., Timothy Fitzgerald, and Jeremy G. Weber. 2016. "Capturing Rents from Natural Resource Abundance: Private Royalties from US Onshore Oil & Gas Production." *Resource and Energy Economics* 46:23–38.

Currie, Janet, Michael Greenstone, and Katherine Meckel. 2017. "Hydraulic Fracturing and Infant Health: New Evidence from Pennsylvania." Science Advances 3 (12): e1603021.

Daly, Herman E. 2005. "Economics in a Full World." *Scientific American* 293 (3): 100–107.

Davis, Charles. 2014. "Substate Federalism and Fracking Policies: Does State Regulatory Authority Trump Local Land Use Autonomy?" *Environmental Science and Technology* 48 (15): 8397–8403. https://doi.org/10.1021/es405095y.

Davis, Charles, and Jonathan M. Fisk. 2014. "Energy Abundance or Environmental Worries? Analyzing Public Support for Fracking in the United States." *Review of Policy Research* 31 (1): 1–16.

Davis, Charles, and Jonathan M. Fisk. 2017. "Mitigating Risks from Fracking-Related Earthquakes: Assessing State Regulatory Decisions." *Society and Natural Resources* 30 (8): 1009–25.

Feyrer, James, Erin T. Mansur, and Bruce Sacerdote. 2017. "Geographic Dispersion of Economic Shocks: Evidence from the Fracking Revolution." *American Economic Review* 107 (4): 1313–34.

Fisk, Jonathan M. 2013. "The Right to Know? State Politics of Fracking Disclosure." *Review of Policy Research* 30 (4): 345–65.

Fisk, Jonathan M. 2016. "Fractured Relationships: Exploring Municipal Defiance in Colorado, Texas, and Ohio." *State and Local Government Review* 48 (2): 75–86.

Fisk, Jonathan M., and A. J. Good. 2019. "Information Booms and Busts: Examining Oil and Gas Disclosure Policies across the States." *Energy Policy* 127:374–81.

Fitzgerald, Timothy. 2013. "Frackonomics: Some Economics of Hydraulic Fracturing." *Case Western Law Review* 63 (4): 1337–62.

Frazier, Reid. 2017. "Pennsylvania Confirms First Fracking-Related Earthquakes." StateImpact, February 18.

Goldstein, Bernard D. 2014. "The Importance of Public Health Agency Independence: Marcellus Shale Gas Drilling in Pennsylvania." *American Journal of Public Health* 104 (2): e13–15.

Goldstein, Bernard D., Elizabeth Ferrell Bjerke, and Jill Kriesky. 2013. "Challenges of Unconventional Shale Gas Development: So What's the Rush." *Notre Dame Journal of Law, Ethics and Public Policy* 27:147–86.

Goldstein, Bernard D., Jill Kriesky, and Barbara Pavliakova. 2012. "Missing from the Table: Role of the Environmental Public Health Community in Governmental Advisory Commissions Related to Marcellus Shale Drilling." *Environmental Health Perspectives* 120 (4): 483–86.

Gullion, Jessica Smartt. 2015. *Fracking the Neighborhood: Reluctant Activists and Natural Gas Drilling.* Cambridge, MA: MIT Press.

Hand, Eric. 2014. "Injection Wells Blamed in Oklahoma Earthquakes." *Science* 345 (6192): 13–14.

Harleman, Max, and Jeremy G. Weber. 2017. "Natural Resource Ownership, Financial Gains, and Governance: The Case of Unconventional Gas Development in the UK and the US." *Energy Policy* 111:281–96.

Hausman, Catherine, and Ryan Kellogg. 2015. "Welfare and Distributional Implications of Shale Gas." *Brookings Papers on Economic Activity* 50 (1): 71–125.

Holahan, Robert, and Gwen Arnold. 2013. "An Institutional Theory of Hydraulic Fracturing Policy." *Ecological Economics* 94:127–34.

Jacobsen, Grant D., and Dominic P. Parker. 2016. "The Economic Aftermath of Resource Booms: Evidence from Boomtowns in the American West." *Economic Journal* 126 (593): 1092–1128.

James, Alexander, and Brock Smith. 2017. "There Will Be Blood: Crime Rates in Shale-Rich US Counties." Journal of Environmental Economics and Management 84:125–52.

Landrigan, Philip J., Howard Frumkin, and Brita E. Lundberg. 2020. "The False Promise of Natural Gas." *New England Journal of Medicine* 382:104–7.

Majumdar, Sarmistha R. 2019. *The Politics of Fracking: Regulatory Policy and Local Community Responses to Environmental Concerns.* New York: Routledge.

Mayfield, Erin N., Jared L. Cohon, Nicholas Z. Muller, Inês M. L. Azevedo, and Allen L. Robinson. 2019a. "Cumulative Environmental and Employment Impacts of the Shale Gas Boom." *Nature Sustainability* 2 (12): 1122–31.

Mayfield, Erin N., Jared L. Cohon, Nicholas Z. Muller, Inês M. L. Azevedo, and Allen L. Robinson. 2019b. "Quantifying the Social Equity State of an Energy System: Environmental and Labor Market Equity of the Shale Gas Boom in Appalachia." *Environmental Research Letters* 14 (12): 124072.

Molotch, Harvey. 1976. "The City as a Growth Machine: Toward a Political Economy of Place." *American Journal of Sociology* 82 (2): 309–32.

Muehlenbachs, Lucija, Elisheba Spiller, and Christopher Timmins. 2015. "The Housing Market Impacts of Shale Gas Development." *American Economic Review* 105 (12): 3633–59.

Murtazashvili, Ilia. 2017. "Institutions and the Shale Boom." *Journal of Institutional Economics* 13 (1): 189–210.

Murtazashvili, Ilia, and Ennio E. Piano. 2018. *The Political Economy of Fracking: Private Property, Polycentricity, and the Shale Revolution.* New York: Routledge.

Murtazashvili, Ilia, and Ennio E. Piano. 2019a. "Governance of Shale Gas Development: Insights from the Bloomington School of Institutional Analysis." *Review of Austrian Economics* 32 (2): 159–79.

Murtazashvili, Ilia, and Ennio E. Piano. 2019b. "More Boon than Bane: How the U.S. Reaped the Rewards and Avoided the Costs of the Shale Boom." *Independent Review* 24 (2).

Raimi, Daniel. 2017. *The Fracking Debate: The Risks, Benefits, and Uncertainties of the Shale Revolution.* New York: Columbia University Press.

US Energy Information Administration. 2016. Maps: Oil and Gas Exploration, Resources and Production. https://www.eia.gov/maps/maps.htm.

Vidic, R. D., S. L. Brantley, J. M. Vandenbossche, D. Yoxtheimer, and J. D. Abad. 2013. "Impact of Shale Gas Development on Regional Water Quality." *Science* 340 (6134).

Vogel, Ronald K., and Bert E. Swanson. 1989. "The Growth Machine versus the Anti-growth Coalition: The Battle for Our Communities." *Urban Affairs Review* 25 (1): 63–85.

Weber, Jeremy G. 2014. "A Decade of Natural Gas Development: The Makings of a Resource Curse?" *Energy and Resource Economics* 37 (1): 168–83.

Weber, Jeremy G., James Wesley Burnett, and Irene M. Xiarchos. 2014. "Shale Gas Development and Housing Values over a Decade: Evidence from the Barnett Shale." USAEE Working Paper No. 14-165. https://ssrn.com/abstract=2467622.

Weber, Jeremy G., Yongsheng Wang, and Maxwell Chomas. 2016. "A Quantitative Description of State-Level Taxation of Oil and Gas Production in the Continental US." *Energy Policy* 96:289–301.

Weitzman, Martin L. 2009. "On Modeling and Interpreting the Economics of Catastrophic Climate Change." *Review of Economics and Statistics* 91 (1): 1–19.

Weitzman, Martin L. 2011. "Fat-Tailed Uncertainty in the Economics of Catastrophic Climate Change." Review of Environmental Economics and Policy 5 (2): 275–92.

Wilber, Tom. 2012. *Under the Surface: Fracking, Fortunes, and the Fate of the Marcellus Shale*. Ithaca, NY: Cornell University Press.

Williamson, Oliver E. 2005. "The Economics of Governance." *American Economic Review* 95 (2): 1–18.

Xu, Minhong, and Yilan Xu. 2020. "Fraccidents: The Impact of Fracking on Road Traffic Deaths." *Journal of Environmental Economics and Management* 101 (102303). https://papers.ssrn.com/sol3/papers.cfm?abstract_id=3175558.

Part I
GOVERNANCE

THE SHALE BOOM IN HISTORICAL PERSPECTIVE

Ilia Murtazashvili and Ennio Piano

News of possible riches spread quickly once drillers in Texas figured out how to profitably frack shale gas. In places like Pennsylvania and West Virginia, communities that had fallen on tough times with the decline of coal and steel production pinned their hopes on shale gas production. In communities that had little experience with natural resource extraction, shale gas seemed like a windfall.

The excitement about the prospects for striking it rich with fracking captivated folks much in the way the US frontier led many to migrate in search of new occupations and opportunities in the nineteenth century. The gold rush in California, which commenced in 1848–1849, was also driven by exuberance around mineral wealth. People flocked to the gold fields in large numbers hoping to strike it rich. Many failed, but the optimism resulted in a veritable boom in gold production.

California's gold rush also conjures images of the Wild West: violence, conflict, and disorder as miners flocked to the gold fields. Perhaps unsurprisingly, journalists have called the shale gas and oil booms a Wild West in places from Ohio (Funk 2011) to North Dakota (Eligon 2013).[1]

Scholars researching the shale boom have also called the shale boom a Wild West in the sense described above. Rabe and Borick (2012) were especially critical of the regulations on drillers during the first years of boom in shale gas production of the Marcellus Shale, arguing that governance was reminiscent of a nineteenth century policy that sought to minimize regulations on energy extraction and to assist development of minerals wherever possible. They argued that

US governments rejected the laissez-faire approach of the nineteenth century but for fracking, regulations were a "blast from the past" because of a perceived lack of regulation. Other descriptions of the shale boom include that it was a "blind rush" (Schmidt 2011) and that we were wading into "untested waters" (Wiseman 2009). Some speculated that the Wild West situation may have been coming to an end as states began adapting earlier legislation governing extraction of conventional minerals to challenges posed by fracking, but still viewed shale gas as unregulated (Revkin 2013).

This chapter offers several points that serve to frame the analysis of governance, planning, and economic impacts associated with fracking in the rest of the book. It shows that some of the accounts of the shale boom were based on perceptions of the past, rather than facts. Our reading of the gold rush history is that it was orderly, and that the government was involved in the governance of miners in California. States have also been responsive to new challenges presented by the shale boom, which calls into question the notion of laissez-faire, or hands-off, regulatory stances by the states. On balance, states that allowed fracking balance economic opportunities with regulation of the socially costly aspects of shale gas production. It is also important to keep in mind that there are many external effects that are not internalized, such as air pollution that contributes to low birth weight and the fact that no amount of fracking is "sustainable." Yet as we will see in the subsequent chapters, there is also a strong case to be made that there ought to be more regulation. Together, the chapters in Part I provide a balanced analysis of governance of the shale boom.

Was the Gold Rush a Wild West?

The American West included all manner of economic activities. One that captures the popular imagination as much as any other is the gold rush. Gold was discovered in California in 1848 on John Sutter's property. Sutter initially claimed land in Mexican California, and after the war with Mexico, the US government agreed to honor those property rights. But word of gold spread quickly, and individuals began to flock to California from around the United States and later from around the world. Popular and historical accounts of the gold rush depict a great deal of crime, absence of institutions, and a very limited role for the state (McKanna 2004, 2002).

The reality of the gold rush is more complicated. One issue is that there was much self-governance during the gold rush. In some areas, there was anarchy, which typically refers to situations in which the government has limited or no control over economic activities (Leeson 2014). Importantly, anarchy—the absence of government telling people what to do and enforcing appropriate behavior—does not always lead to disorder. Self-governance often works well in regulating human behavior (Anderson and Hill 2004).

Individuals often devised their own institutions during the gold rush, which emerged quickly in response to new economic opportunities. The first institutions governing gold extraction were simple and small. These mining camps, which consisted of teams of five to ten men, allocated access rights to new lodes on a first-possession basis (the first collection to find gold had access rights to a small area), with labor effort and profits shared among the group's members (Zerbe and Anderson 2001).

The mining camps showed that people could cooperate in the absence of law, but they were unable to impose order on the gold fields as miners flocked there in search of riches (Umbeck 1977). Eventually, mining colonies consisted of dozens and even hundreds of camps and so miners established a new form of governance: the mining district, which was established by popular consent among the individual miners and mining camps. It was in effect a constitution to govern individuals and mining camps in the region. The mining district became the basis for law and order among oftentimes hundreds of mining camps. These districts were governance organizations with executives, a collective decision-making body, and even courts to adjudicate disputes among miners (Murtazashvili 2013). Generally, the districts had both courts for the miners and what they called the "people's court." The latter were the courts for crimes more generally and had more of a feature of vigilante justice.[2]

Although the gold rush has been popular as an explanation for the spontaneous emergence of private property, Clay and Wright (2005) showed that property rules during the gold rush were less a private property system than a loose collection of norms for managing access to a common-pool resource. Miners had to physically occupy their claims or else they lost their "rights" to the claim. That is, the mining codes legitimized claim jumping, which forced individuals to continuously occupy their property. A situation in which people have to continually stand guard over their possessions signifies weak private property rights (Bromley 2006). There is also little evidence that miners made much money, as most ended up poor, although the surrounding areas gained (Clay and Jones 2008). It is also unclear whether popular tribunals were effective in dispensing "justice." The decisions of frontier courts were often viewed as quite arbitrary, although there was also a great deal of order in these tribunals (McDowell 2004).

What we ought to take from this is that the miners recognized a need for governance, first with the mining camps and then with the mining districts. It was of course imperfect, but still a system of governance. The federal government made possible the gold rush by starting a war with Mexico that opened California to settlement. Once squatters flocked to the gold fields, Congress attempted to regulate economic activities. The Land Claims Act of 1851 established tribunals to sort out competing claims to ownership. It was far from perfect, but it was an effective initial attempt to regulate economic activities in an uncertain environment (Clay 1999). Congress responded quickly to the challenges presented by the mining boom with the Mining Act of 1866, which recognized the rules that emerged through a spontaneous process and appeared to work well for the miners (Libecap 1989). Although it may seem that there was a long delay in institutional and legal reform, consider the regulation of the open range as another example. Open range ranching emerged in the 1860s, but Congress did not reform the open range system until the Taylor Grazing Act of 1934. Congress was quick to regulate mines in comparison to the range cattle industry.

This brief discussion of the gold rush has important implications. One is that there was governance, initially informal and then formal. The mining districts were an early effort to plan for the gold boom. The government initially attempted to sort out competing claims, and later established a legal framework that would govern mining to this day. Thus, the image of the gold rush as a Wild West does not exactly fit the reality of gold mining in its first decades. As we will see, the metaphor is not always apt for the shale boom.

The Image and Reality of the Shale Boom

Calling the shale boom a Wild West is at its core a critique of governance of the shale boom. Below, we suggest several aspects of this critique. To be sure, there are many other ways to characterize governance of fracking. What follows is a set of reasonable criticisms. Yet for each, there is more order and much more regulation than suggested by a Wild West interpretation of the shale rush.

Is Fracking a New and Untested Technology?

One criticism of the shale boom is that the "technology" is new and untested (Wiseman 2009). To be sure, the shale boom is a story of innovation (Golden and Wiseman 2015). Yet the novelty of fracking technology is an overstatement. Fracking combines two techniques—horizontal drilling and hydraulic fracturing—that have been in use for half a century. The novel aspect of fracking is the use of chemicals in the fracturing process. These chemicals, which make the water used in the fracking machines slicker and hence more effective in the process of mining natural gas, create different challenges than with conventional oil and gas, namely, a nonpoint pollution externality from groundwater extraction.

Conventional oil and gas extraction, in contrast, presents a challenge of a pumping race that is largely irrelevant to fracking (Holahan and Arnold 2013).

The disclosure of chemicals used in the fracturing process is the area where regulation might be lacking. There has been substantial debate about what companies have to disclose, as well as when to disclose it (Fisk and Good 2019, Fisk 2013). For example, some companies only wanted to disclose information about the chemicals used in the fracturing process in the event of a disaster. Thus, while novelty is overstated, the lack of information about chemicals involved could reasonably be described as a Wild West situation in the sense that the chemicals released were not always disclosed, but as the research above indicates, the states were by no means powerless to regulate disclosure. To the contrary, all states have policies, though their effectiveness is sometimes questionable, and the effectiveness of those regulations may be diminished by lobbying by business.

Was the Land Rush Chaotic?

There are many examples in US history of land rushes that resulted in conflict and dissipation of resources, such as the Oklahoma Land Rush of 1893 (Allen and Leonard 2020). The shale boom, because it requires landmen to sign leases before companies commence fracking, has also been described as a land rush. Yet this land rush differed from previous ones. First and foremost, the rush was to sign leases, rather than to allocate land to settlers. The property rights themselves—the rights to minerals that lie below the surface of land—were established long ago. Indeed, one of the features of the US shale boom was that it occurred in a fairly well-established property rights framework (Murtazashvili 2017).

The leasing process usually involved the following. After gas companies identify the location of shale gas, they employ the service of landmen to locate who owns mineral rights to land and negotiate with owners of mineral rights. Property rights are bundles of rights that can be divided and subdivided (Ellickson 1993). In the United States, the surface estate (land) is often separated from the mineral estate. When a contract is successfully negotiated, it concludes with an agreement over an up-front bonus payment royalty. Most states have a minimum royalty requirement (typically about 12.5 percent of the profits from shale gas extraction) that alleviates some of the distributive questions associated with shale production.

The property regime is not without controversies. It is common in state law that the mineral estate dominates, which means that surface owners who do not have mineral rights have to allow reasonable access for the mineral rights owners to develop the mineral estate. Recent work finds that there is greater dissatisfaction with fracking on split estates in comparison to situations when an individual owns both the surface and mineral estate (Collins and Nkansah 2015). Federal ownership of the mineral estates may also be a source of transaction costs that undermine the value of mineral rights leases (Fitzgerald 2010). Moreover, the leasing process may disadvantage minorities (Vissing 2015). Forced pooling, which requires people to participate in drilling when the land surrounding them has been leased, is sometimes opposed by property owners who may object to royalty payments imposed by the group, or because they may want a better deal in terms of their lease. They may also oppose drilling out of concern about environmental consequences.

The landmen, depicted in a Hollywood account in *Promised Land* and documentaries such as *Gasland*, are often viewed as unscrupulous. While the landmen have an interest in getting the best deal for gas companies, they do not operate in an institutional vacuum. In addition to the governance of their own professional associations, there are statutory requirements regarding royalties.[3] Indeed, fraud has been documented. Another issue is that landmen may use information to their advantage. Black, McCoy, and Weber (2018) find that there is some evidence that leases are not efficient, as values do not move as much as they should with differences in shale production. In economics, this is called pass-through: when production is strong, leases should be more valuable. When lease values do not change all that much with the market, it suggests something is amiss.

A more useful way to think of landmen is as people providing a service, or as entrepreneurs. The reason they exist is because the law required that the companies contract with the owners of mineral rights even if they did not know they own the mineral rights. There are severe penalties for fraud, which is a federal offense since the interstate nature of the industry means that any duplicity could end up with a federal case—and federal prison. They also have their own codes of conduct. Even if that does not work, lawyers often organize landowners, combining them into a sellers' pool that allows them to improve their bargaining position with landmen. These sellers' pools can be thought of as improving the bargaining position of mineral rights owners, including providing information on the "optimal" lease contract (Murtazashvili and Piano 2018). Thus, while one might make a case more regulations are needed, or better ones, there is not much evidence that the contracting process is a Wild West.

Did Federalism Result in Fractured, Ineffective Regulation?

One of the most important governance questions is whether regulators responded effectively to the shale boom. This question can only be understood in the context of federalism. For Vincent Ostrom (1994), federalism is the foundation of a self-governing society. Its defining feature is that governance authority is shared at multiple levels (Bednar 2008).

In the context of fracking, the authority to regulate shale gas is shared among the states, local governments, and the federal government. States have the greatest regulatory authority over fracking (Spence 2013, Arnold and Holahan 2014). The reason is that states have traditionally had the authority to regulate for the purposes of health, safety, and welfare. Local governments share authority to regulate fracking through zoning, which is a critical source of regulatory authority. The federal government oversees fracking, with the federal Environmental Protection Agency coordinating research on hydraulic fracturing. In addition, Congress could likely regulate hydraulic fracturing because of its potential impact on groundwater.

The federalism of fracking has been criticized because it results in a fragmented regulatory regime (Warner and Shapiro 2013, Wiseman 2010). Indeed, the states have responded with different regulations: the response has ranged from fracking, full speed ahead, to complete bans on development (Richardson et al. 2013). Yet it is not clear that divergence of regulations should be taken as a failure of governance.

A comparison of Pennsylvania and New York is a good example. Pennsylvania's Act 13, which was signed into law in 2012, denied local governments the authority to prohibit hydraulic fracturing. This provision resulted in several municipalities challenging the law in court because, they argued, it undermined the principles of self-governance.[4]

To be sure, Act 13 was a constraint on local autonomy, but there was a compelling reason for it. Part of the reason for the law is that Pennsylvania has over 2,500 municipal governments. It is challenging for business to operate in an environment in which each of the local governments had the ability to choose its own regulatory framework. In addition, many of the local governments did not have a dedicated planning or zoning board, and many of them did not have much in the way of resources. In this context, the Act 13 provisions, which established for communities a model zoning ordinance, were a solution to a governance challenge. It was paternalistic in that it involved the state telling communities what to do. However, it did not deny authority to determine what kind of economic development is appropriate in a community. Each community could still use zoning to determine if *development* was allowable in parts of their community; they just could not create blanket regulations governing *fracking*.

Around the same time leases were being signed in Pennsylvania, landmen were making deals with mineral rights owners in New York. Yet while production commenced quickly and soon skyrocketed in Pennsylvania after leases were signed, the New York government initiated a study that was a de facto moratorium. It was still in place a decade and a half after the first wells were drilled in its neighboring state; as of April 2020, Governor Andrew Cuomo has made the ban permanent.

How one interprets these responses depends in part on how one views federalism. One of the defining features of decentralized, polycentric governance is that there will be variation in regulatory responses for a variety of reasons besides environmental considerations, such as some jurisdictions may value economic growth over environmental considerations, or some jurisdictions may want to experiment with new regulations (E. Ostrom 2005, Rodden 2006). Experimentation may also result in states adopting similar policies once they see what works (Shipan and Volden 2006).

Another reason for the diversity of responses is differences in preferences for economic development. Some communities place more value on the importance of economic development. Indeed, such variation is perhaps the most important economic justification for why public goods, including regulation, ought to be provided locally, rather than by the national government. When people can vote with their feet, voter mobility is likely to induce a more efficient supply of public goods (Tiebout 1956, Somin 2011). Thus, what is optimal from a regulatory perspective may differ even if the extent of externalities is the same.

The EPA, while not the lead regulatory agency, is far from a passive player in the shale boom. The EPA has commissioned many scientific studies and collected a great deal of anecdotal evidence regarding shale. The EPA stands ready to regulate in the event there are spillovers that affect more than one state, or if there is an environmental externality that requires more in the way of federal regulation. It also exerts regulatory control over air pollution, which is increasingly recognized as the most fundamental type of pollution from fracking (Mayfield et al. 2019). The global commons aspect of pollution often benefits from higher-level imposition of regulation to internalize externalities. There is also a collective action problem confronting state governments: the first states to regulate fracking may be at a competitive disadvantage. Federal regulation can overcome these commons problems by imposing uniform standards. Even so, cities remain an important source of public policies to address pollution, and so there remains an important justification for federalism even though air pollution, by its nature, contributes to a global commons problem (E. Ostrom 2010b).

Are There Uncontrolled Externalities with Shale Gas Development?

A more specific question concerns the extent that regulations address the externalities associated with fracking. To consider this issue, it is useful to first note the differences between conventional and unconventional oil and gas extraction. Oil and gas are examples of common-pool, finite resources. According to Elinor Ostrom, who in 2009 was recognized with the Nobel Prize in Economics for analyzing these types of resources, the defining features of common-pool resources are that they are nonexcludable and divisible. Nonexcludability refers to the challenge of establishing property rights to the resource, while divisibility means it gets used up (E. Ostrom 2010a).

The economic problem with conventional oil and gas arises when many people may have access to a reservoir. This results from the location of conventional oil and gas: they are easy to tap from the surface, and the minerals are not housed in a semipermeable rock that oftentimes lies more than a mile beneath the earth's surface. In such cases, individuals have incentives to race to establish property rights because they cannot exclude others from the reservoir (Anderson and Hill 1990). Hardin (1968) referred to such situations as a tragedy of the commons, which arises when private decisions lead to collectively costly outcomes. Fortunately, the tragedy of the commons is not inevitable (Frischmann, Marciano, and Ramello 2019). The underlying challenge with overextraction of a common-pool resource is to establish property rights over the object of value. Property rights are the rules governing ownership in society (Barzel 1997). They include private property, common property, and state ownership (Bromley 1991). The absence of property rights is open access. The tragedy of the commons results from open access; the solution is to create an appropriate property regime (Cole, Epstein, and McGinnis 2014).

The institutional solutions to the problem of competition to extract oil or gas from a well as quickly as possible are unitization and well-spacing agreements. Unitization is an institutional arrangement that creates a collective property right for all who have access to a reservoir. Hence, all are residual claimants in the long-run production of the reservoir. Such unitization agreements overcome the tragedy of the commons with conventional oil and gas extraction (Libecap and Smith 2002).

The impermeability and location of shale gas far below the earth's surface mean that racing is not as much of a challenge for extraction. However, there are many other challenges from shale gas development (Sovacool 2014). Rather than review the literature, we make a few observations. First, it is extremely challenging to identify a causal impact of fracking on surface water or groundwater (Vidic et al. 2013; Olmstead et al. 2013). Although most studies do not find clear evidence of groundwater contamination, there are nontrivial risks (Mauter et al. 2014). There is also substantial uncertainty about the adverse consequences of methane for global warming, with several scientists suggesting that the early estimates of the adverse consequences are dramatically overstating the harms

(Cathles, III et al. 2012). In addition, while there is evidence that fracking affects property values, these studies show that it is *perception* of groundwater contamination that influences prices (Muehlenbachs, Spiller, and Timmins 2015). It is important to keep in mind that the health costs, including from pollution, have less to do with fracking than with use of diesel fuel in the process (Raimi 2017).

These studies affirm what is true of all economic development: mineral extraction has social costs. The relevant question is then whether regulation addresses them. What we know is that each state has a formidable regulatory apparatus. For example, the setback provisions of Pennsylvania's Act 13 are designed to address the potential that fracking fluid contaminates groundwater. In addition, there are techniques to deal with fugitive methane that are standard in the industry, such as "flaring"—burning gas at the wellhead when it escapes. Flaring mitigates the challenge with fugitive methane.

There remain challenges from rent-seeking by business, which may lead to opposition to disclosure laws and other regulations that are important to addressing externalities. It is also clear that there are external effects from air pollution locally and globally. There are serious adverse health effects from air pollution when people live very close to a well (Currie, Greenstone, and Meckel 2017), as well as concerns that fracking increases household concentrations of radon (Black, McCoy, and Weber 2019).

The proof with regulation is in the pudding. Our previous review of the literature finds that the benefits far exceed the costs of shale production (Murtazashvili and Piano 2019b). Studies that account for the costs of fracking on the environment have found substantial net benefits (Hausman and Kellogg 2015), though some find net costs when accounting for the impact of pollution on human health (Mayfield et al. 2019). Thus, there is at least a plausible case to be made that there is regulation that accounts for the external effects of shale gas development, though a case can also be made that pollution especially remains a challenge with shale gas—though the effect of pollution is less a problem with fracking per se than a consequence of use of oil-fueled machines in the process of extracting minerals from the earth.

Is Government Absent?

Perhaps more than anything else, the image of the Wild West in popular imagination is one of absentee government, or anarchy. Earlier, we explained this view is not especially accurate: the federal government had a significant influence during the gold rush, providing a legal framework to govern the development of the frontier and to resolve conflicts over land, as did the self-governing associations the miners established themselves.

The states have certainly been active in regulating fracking, as well as providing an institutional framework to govern shale gas development. Government is the ultimate source of property rights, which we have seen are clearly established. Governments have also taken on the role of development planners, presiding over and even facilitating shale gas development. One of the important examples is Act 13, though all states that allow fracking had regulations in place to govern it early on in the shale boom (Davis 2012). Moreover, governments at all levels have been active regulating shale gas, including downstream uses. For example, as of 2019, the city of Pittsburgh has a moratorium on new "cracker" plants that use by-products from fracking to make plastics.

The most obvious extension of state power to regulate is the de facto prohibition of shale gas development in New York. The government was woefully unable to keep squatters off its land during the nineteenth century. Indeed, much of the land that the squatters colonized during the gold rush was privately owned, including John Sutter's land. Sutter's Mill was private property and the federal government could not do much to prevent anyone from carrying off gold he thought was his. It is not even conceivable that the government in the nineteenth century could prohibit free mining. Today, the state governments not only have the authority to prevent development of minerals, but have done so even after gas companies, landmen, and owners signed lease contracts. The fact that the state of New York has prevented fracking after a decade and a half of booming shale gas production in Pennsylvania shows that the states are exceptionally powerful as regulatory bodies, which includes their power to entirely prohibit activities.

The West Wasn't Wild, and Neither Was the Shale Boom

Mineral extraction in the nineteenth century was not as wild as it is often depicted, and neither was the shale boom. The reality of the gold rush during the nineteenth century was the rapid emergence of informal institutions to provide order, as well as a significant role for the state. Regulations emerged rather quickly to govern the gold fields.

Our view is that the metaphor of the Wild West is misleading with fracking. The process of contracting for property rights occurred quickly. Regulators responded quickly by modifying the framework of conventional oil and gas law to deal with challenges posed by fracking. Institutions and policies made possible the rapid expansion of economic activities by providing the basic institutional structure that permitted the expansion of a new sector. The states have proven more than capable of regulating the shale boom, illustrating the continued benefits of polycentric governance of fracking (Murtazashvili and Piano 2018).

Of course, it is reasonable to question the diversity of regulatory responses. Should we worry that some states frack away, while others banned it? Such differences are expected in the US federal system. Governments at all levels also confront significant information problems regarding how to regulate new activities, especially in regard to economic activities with environmental consequences (Pennington 2008, 2011). One of the virtues of federalism is that it encourages experimentation in response to new challenges. Regulatory diversity can be viewed as a strength in a context of uncertainty about the consequences of shale gas development (Murtazashvili and Piano 2019a).

Our goal in this chapter has been to deal with the "big picture" comparisons. But the devil is in the details when it comes to understanding a phenomenon as complex as the shale boom. The chapters ahead offer additional evidence regarding the claims we considered in this chapter. Many of them are to some extent at odds with our interpretation of the shale boom. It is our hope that this chapter adds to the robust debate about the governance of shale gas.

Notes

1. North Dakota has seen a rapid increase in drilling for shale oil, rather than gas. Fracking is a common technique used to extract oil and gas from shale, although the focus in this chapter is on fracking for gas. See chapter 6 for a case study on Williston, ND, and its boom-bust outcomes in planning.

2. In fact, vigilantism was a form of expertise on the frontier that often contributed to social order (Obert 2014; Obert and Mattiacci 2018).

3. See chapter 3 for an extended discussion on the role of landmen throughout West Virginia's resource history.

4. For a more thorough discussion of Act 13, see chapter 2. See also Murtazashvili (2015).

References

Allen, Douglas W., and Bryan Leonard. 2020. "Rationing by Racing and the Oklahoma Land Rushes." *Journal of Institutional Economics* 16 (2): 127–44.

Anderson, Terry L., and Peter J. Hill. 1990. "The Race for Property Rights." *Journal of Law and Economics* 33 (1): 177–97.

Anderson, Terry L., and Peter J. Hill. 2004. *The Not So Wild, Wild West: Property Rights on the Frontier.* Palo Alto, CA: Stanford University Press.

Arnold, Gwen, and Robert Holahan. 2014. "The Federalism of Fracking: How the Locus of Policy-Making Authority Affects Civic Engagement." *Publius* 44 (2): 344–68.

Barzel, Yoram. 1997. *Economic Analysis of Property Rights*. New York: Cambridge University Press.

Bednar, Jenna. 2008. *The Robust Federation*. New York: Cambridge University Press.

Black, Katie Jo, Shawn J. McCoy, and Jeremy G. Weber. 2018. "When Externalities Are Taxed: The Effects and Incidence of Pennsylvania's Impact Fee on Shale Gas Wells." *Journal of the Association of Environmental and Resource Economists* 5 (1): 107–53.

Black, Katie Jo, Shawn J. McCoy, and Jeremy G. Weber. 2019. "Fracking and Indoor Radon: Spurious Correlation or Cause for Concern?" Journal of Environmental Economics and Management 96:255–73.

Bromley, Daniel W. 1991. *Environment and Economy: Property Rights and Public Policy*. Oxford: Blackwell Publishers.

Bromley, Daniel W. 2006. *Sufficient Reason: Volitional Pragmatism and the Meaning of Economic Institutions*. Princeton: Princeton University Press.

Cathles, Lawrence M., III, Larry Brown, Milton Taam, and Andrew Hunter. 2012. "A Commentary on 'The Greenhouse-Gas Footprint of Natural Gas in Shale Formations' by R. W. Howarth, R. Santoro, and Anthony Ingraffea." *Climatic Change* 113 (2): 525–35.

Clay, Karen. 1999. "Property Rights and Institutions: Congress and the California Land Act 1851." *Journal of Economic History* 59 (1): 122–42.

Clay, Karen, and Randall Jones. 2008. "Migrating to Riches? Evidence from the California Gold Rush." *Journal of Economic History* 68 (4): 997.

Clay, Karen, and Gavin Wright. 2005. "Order without Law? Property Rights during the California Gold Rush." *Explorations in Economic History* 42 (2): 155–83.

Cole, Daniel H., Graham Epstein, and Michael D. McGinnis. 2014. "Digging Deeper into Hardin's Pasture: The Complex Institutional Structure of 'The Tragedy of the Commons.'" *Journal of Institutional Economics* 10 (3): 353–69.

Collins, Alan R., and Kofi Nkansah. 2015. "Divided Rights, Expanded Conflict: Split Estate Impacts on Surface Owner Perceptions of Shale Gas Drilling." *Land Economics* 91 (4): 688–703.

Currie, Janet, Michael Greenstone, and Katherine Meckel. 2017. "Hydraulic Fracturing and Infant Health: New Evidence from Pennsylvania." Science Advances 3 (12): e1603021.

Davis, Charles. 2012. "The Politics of 'Fracking': Regulating Natural Gas Drilling Practices in Colorado and Texas." *Review of Policy Research* 29 (2): 177–91.

Eligon, John. 2013. "An Oil Town Where Men Are Many, and Women Are Hounded." *New York Times*, January 15.

Ellickson, Robert C. 1993. "Property in Land." *Yale Law Journal* 102:1315–1400.

Fisk, Jonathan M. 2013. "The Right to Know? State Politics of Fracking Disclosure." *Review of Policy Research* 30 (4): 345–65.

Fisk, Jonathan M., and A. J. Good. 2019. "Information Booms and Busts: Examining Oil and Gas Disclosure Policies across the States." *Energy Policy* 127:374–81.

Fitzgerald, Timothy. 2010. "Evaluating Split Estates in Oil and Gas Leasing." *Land Economics* 86 (2): 294–312.

Frischmann, Brett M., Alain Marciano, and Giovanni Battista Ramello. 2019. "Retrospectives: Tragedy of the Commons after 50 Years." *Journal of Economic Perspectives* 33 (4): 211–28.

Funk, John. 2011. "Rural Ohio Is the Wild West as Gas and Oil Companies Compete for Drilling Rights." *Cleveland Plain Dealer*, October 22.

Golden, John M., and Hannah Jacobs Wiseman. 2015. "The Fracking Revolution: Shale Gas as a Case Study in Innovation Policy." *Emory Law Journal* 64 (4): 955–1040.

Hardin, Garrett. 1968. "The Tragedy of the Commons." *Science* 162 (3859): 1243–48.

Hausman, Catherine, and Ryan Kellogg. 2015. "Welfare and Distributional Implications of Shale Gas." *Brookings Papers on Economic Activity* 50 (1): 71–125.

Holahan, Robert, and Gwen Arnold. 2013. "An Institutional Theory of Hydraulic Fracturing Policy." *Ecological Economics* 94:127–34.

Leeson, Peter T. 2014. *Anarchy Unbound: Why Self-Governance Works Better Than You Think*. New York: Cambridge University Press.

Libecap, Gary D. 1989. *Contracting for Property Rights*. New York: Cambridge University Press.

Libecap, Gary D., and James L. Smith. 2002. "The Economic Evolution of Petroleum Property Rights in the United States." *Journal of Legal Studies* 31 (S2): S589–608.

Mauter, Meagan S., Pedro J. J. Alvarez, Allen Burton, Diego C. Cafaro, Wei Chen, Kelvin B. Gregory, Guibin Jiang, Qilin Li, Jamie Pittock, and Danny Reible. 2014. "Regional Variation in Water-Related Impacts of Shale Gas Development and Implications for Emerging International Plays." *Environmental Science and Technology* 48 (15): 8298–8306.

Mayfield, Erin N., Jared L. Cohon, Nicholas Z. Muller, Inês M. L. Azevedo, and Allen L. Robinson. 2019. "Cumulative Environmental and Employment Impacts of the Shale Gas Boom." *Nature Sustainability* 2 (12): 1122–31.

McDowell, Andrea. 2004. "Real Property, Spontaneous Order, and Norms in the Gold Mines." *Law and Social Inquiry* 29 (4): 771–818.

McKanna, Clare V. 2002. *Race and Homicide in Nineteenth-Century California*. Reno: University of Nevada Press.

McKanna, Clare V. 2004. "Enclaves of Violence in Nineteenth-Century California." *Pacific Historical Review* 73 (3): 391–424.

Muehlenbachs, Lucija, Elisheba Spiller, and Christopher Timmins. 2015. "The Housing Market Impacts of Shale Gas Development." *American Economic Review* 105 (12): 3633–59.

Murtazashvili, Ilia. 2013. *The Political Economy of the American Frontier*. New York: Cambridge University Press.

Murtazashvili, Ilia. 2015. "Origins and Consequences of State-Level Variation in Shale Regulation: The Cases of Pennsylvania and New York." In *Economics of Unconventional Shale Gas Development*, edited by Yongsheng Wang and William Hefley, 179–201. Cham: Springer.

Murtazashvili, Ilia. 2017. "Institutions and the Shale Boom." *Journal of Institutional Economics* 13 (1): 189–210.

Murtazashvili, Ilia, and Ennio E. Piano. 2018. *The Political Economy of Fracking: Private Property, Polycentricity, and the Shale Revolution*. New York: Routledge.

Murtazashvili, Ilia, and Ennio E. Piano. 2019a. "Governance of Shale Gas Development: Insights from the Bloomington School of Institutional Analysis." *Review of Austrian Economics* 32 (2): 159–79.

Murtazashvili, Ilia, and Ennio E. Piano. 2019b. "More Boon than Bane: How the U.S. Reaped the Rewards and Avoided the Costs of the Shale Boom." *Independent Review* 24 (2).

Obert, Jonathan. 2014. "The Six-Shooter Marketplace: 19th-Century Gunfighting as Violence Expertise." *Studies in American Political Development* 28 (1): 49–79.

Obert, Jonathan, and Eleonora Mattiacci. 2018. "Keeping Vigil: The Emergence of Vigilance Committees in Pre-Civil War America." *Perspectives on Politics* 16 (3): 600–616.

Olmstead, Sheila M., Lucija A. Muehlenbachs, Jhih-Shyang Shih, Ziyan Chu, and Alan J. Krupnick. 2013. "Shale Gas Development Impacts on Surface Water Quality in Pennsylvania." *Proceedings of the National Academy of Sciences* 110 (13): 4962–67.

Ostrom, Elinor. 2005. *Understanding Institutional Diversity*. Princeton: Princeton University Press.

Ostom, Elinor. 2010a. "Beyond Markets and States: Polycentric Governance of Complex Economic Systems." *American Economic Review* 100 (3): 641–72.

Ostrom, Elinor. 2010b. "Polycentric Systems for Coping with Collective Action and Global Environmental Change." *Global Environmental Change* 20 (4): 550–57.

Ostrom, Vincent. 1994. *The Meaning of American Federalism: Constituting a Self-Governing Society*. San Francisco: Institute for Contemporary Studies.

Pennington, Mark. 2008. "Classical Liberalism and Ecological Rationality: The Case for Polycentric Environmental Law." *Environmental Politics* 17 (3): 431–48.

Pennington, Mark. 2011. *Robust Political Economy: Classic Liberalism and the Future of Public Policy*. Cheltenham: Edward Elgar Publishing.

Rabe, Barry G., and Christopher P. Borick. 2012. "Gas Drillers' New Wild West." *Philadelphia Inquirer*, February 16.

Raimi, Daniel. 2017. *The Fracking Debate: The Risks, Benefits, and Uncertainties of the Shale Revolution*. New York: Columbia University Press.

Revkin, Andrew C. 2013. "Is 'Wild West' Era for Gas Drilling Coming to an End?" *New York Times*, January 17.

Richardson, Nathan, Madeline Gottlieb, Alan Krupnick, and Hannah J. Wiseman. 2013. "The State of State Shale Gas Regulation." Resources for the Future Report. https://media.rff.org/archive/files/sharepoint/WorkImages/Download/RFF-Rpt-StateofStateRegs_Report.pdf.

Rodden, Jonathan. 2006. *Hamilton's Paradox: The Promise and Peril of Fiscal Federalism*. Cambridge: Cambridge University Press.

Schmidt, Charles W. 2011. "Blind Rush? Shale Gas Boom Proceeds amid Human Health Questions." *Environmental Health Perspectives* 119 (8): a348.

Shipan, Charles R., and Craig Volden. 2006. "Bottom-Up Federalism: The Diffusion of Antismoking Policies from US Cities to States." *American Journal of Political Science* 50 (4): 825–43.

Somin, Ilya. 2011. "Foot Voting, Political Ignorance, and Constitutional Design." *Social Philosophy and Policy* 28 (1): 202–27.

Sovacool, Benjamin K. 2014. "Cornucopia or Curse? Reviewing the Costs and Benefits of Shale Gas Hydraulic Fracturing (Fracking)." *Renewable and Sustainable Energy Reviews* 37:249–64.

Spence, David B. 2013. "Backyard Politics, National Policies: Understanding the Opportunity Costs of National Fracking Bans." *Yale Journal of Regulation* 30:30A–475.

Tiebout, Charles M. 1956. "A Pure Theory of Local Expenditures." *Journal of Political Economy* 64 (5): 416–24.

Umbeck, John. 1977. "A Theory of Contract Choice and the California Gold Rush." *Journal of Law and Economics* 20:421.

Vidic, R. D., S. L. Brantley, J. M. Vandenbossche, D. Yoxtheimer, and J. D. Abad. 2013. "Impact of Shale Gas Development on Regional Water Quality." *Science* 340 (6134).

Vissing, Ashley. 2015. "Private Contracts as Regulation: A Study of Private Lease Negotiations Using the Texas Natural Gas Industry." *Agricultural and Resource Economics Review* 44 (2): 120–37.

Warner, Barbara, and Jennifer Shapiro. 2013. "Fractured, Fragmented Federalism: A Study in Fracking Regulatory Policy." *Publius: The Journal of Federalism* 43 (3): 474–96.

Wiseman, Hannah J. 2009. "Untested Waters: The Rise of Hydraulic Fracturing in Oil and Gas Production and the Need to Revisit Regulation." *Fordham Environmental Law Review* 20:115–71.

Wiseman, Hannah J. 2010. "Regulatory Adaptation in Fractured Appalachia." *Villanova Environmental Law Journal* 21:229.

Zerbe, Richard O., and C. Leigh Anderson. 2001. "Culture and Fairness in the Development of Institutions in the California Gold Fields." *Journal of Economic History* 61 (1): 114–43.

LOCAL JURISDICTIONS AND VARIATIONS IN STATE LAW IN THE MARCELLUS SHALE REGION

Heidi Gorovitz Robertson

Citizens, businesses, and state and local governments are interested in shale oil and gas related decision-making. Some encourage drilling—believing it will bring jobs, an influx of related business, and prosperity. Others oppose shale development because they believe its reliance on hydraulic fracturing technology is dangerous to the water supply, air quality, health, and the social and economic structures of their communities. They want their local governments to ban, control, or influence it. Local governments, therefore, want the authority to decide whether shale oil and gas development will occur within their jurisdictions; and if so, where and under what circumstances these activities will occur.

The regulatory actions a local government may take are defined by the broader legal systems in which the community is situated, and there are important differences among state legal systems. The ability of local governments in Pennsylvania to make decisions regarding local drilling differs from those in Ohio, New York, or elsewhere. Local governments may be limited by their state constitutions, state statutes, and the interpretations of those documents by their state courts.

This chapter will consider each of those state-level legal factors—constitutions, statutes, and judicial interpretations—within the Marcellus region, as well as their impacts on local governments' shale oil and gas related decision-making. Through a comparison of critical cases, it shows that local governments should have a role in determining whether, where, and how drilling happens within local boundaries. Moreover, the analysis illustrates the value of a polycentric perspective on shale gas development in which multiple authorities influence how local governments respond to challenges and opportunities presented by shale gas development.

Primary Factors in State Legal Regimes

There are three interconnected sources of law that control the ability of local jurisdictions to control or influence shale oil and gas development: state constitutions, state legislation, and state courts. State constitutions vary in both language and interpretation. For example, some constitutions include "home rule" protection for local jurisdictions. "Home rule" theoretically protects local governments' power to decide how their communities operate. However, some state courts, in particular Ohio's, interpret home rule protection narrowly, as essentially applying only to decisions regarding the structure and operation of local government. Other states' courts interpret similar language more broadly, giving local governments much more leeway in the actions they can take under the auspices of home rule.

Beyond variations in and interpretations of home rule language, state constitutions may include other provisions that affect the authority of local governments. For example, some state constitutions, like Pennsylvania's, require governmental authorities to act in the best interest of the environment. This would, of course, affect what a local government (or a state agency) could do.

Aside from the federal environmental laws on air, water, and waste management, there is little federal law governing shale oil and gas development. The vast majority of applicable law comes from state legislatures and the state agencies implementing state law. As expected, the several Marcellus states chose their own terms for governing shale oil and gas development. There is substantial consistency regarding which shale-related activities states choose to regulate, for example, the location and spacing of wells, methods of drilling, disposal of oil and gas wastes, and site restoration (Richardson et al. 2013; Wiseman 2014). These activities concern how and where shale oil and gas will be developed, and they are the precise issues local jurisdictions seek to influence.

Although states regulate similar types of activities, they differ in *how* and the *extent* to which they regulate (Richardson et al. 2013; Wiseman 2014). The manner and extent of regulation varies because states differ from one another in several relevant ways—in citizens' perceptions of environmental risks, in actual environmental risks resulting from their geography and geology, in their political environments, economic concerns, history, and more. The actions that local authorities may take depend on state legislation; some states' legislation preempts local control of oil and gas activities because the state government is the regulating authority instead. Local ordinances may not conflict with state law; and if ordinances purport to regulate in areas reserved for the state, state courts will void them.

State legislatures also determine which state agency will control oil and gas development regulation. It matters whether the legislature delegates regulatory authority to an environmental agency, a natural resources agency, or a combination of the two. The missions of these agencies differ in important ways, as can the extent to which they involve local governments in their administrative processes. It matters to local governments, therefore, which type of agency the legislature assigns regulatory responsibility.

By interpreting state oil and gas laws, state courts help determine the decision-making authority of local governments. They interpret statutory provisions on preemption and home rule, and state constitutions' provisions on home rule, environmental protection, and anything else relevant. When state oil and gas laws seek to preempt local control, courts determine the extent of the preemption. State supreme courts are political creatures, generally populated by elected justices, so they vary in their interpretations of statutory language and constitutions even when that language may be similar or even identical.

Variations in Local Control in the Marcellus Shale Region

This section describes variations in state oil and gas laws that seek to preempt local regulation of shale oil and gas development in Ohio, Pennsylvania, and New York. It describes how state courts have interpreted these laws within the context of their state constitutions.

Ohio

Ohio governor John Kasich sought to encourage the state's shale oil and gas industry believing it to be good for Ohio's economy. To do this, he said Ohio needed a regulatory system that was uniform, statewide. Uniformity would preclude local ordinances because those ordinances would create local variations in the rules, complicating matters for the incoming industry. Ohio's original oil and gas law was silent regarding local regulatory authority and silent on any delegation of authority to a specific state agency. In 2010, the Ohio legislature amended the oil and gas statute to create the unified system the governor sought. It gave "sole and exclusive authority" over the location, spacing, and permitting of oil and gas wells to the Ohio Department of Natural Resources' (DNR) Division of Oil and Gas Resources Management. Then, over a two-year period the Ohio legislature serially amended the law to expand the enumerated activities over which the Ohio DNR has "sole and exclusive authority" and to explicitly preempt local control.

Despite this preemption, local governments sought to influence oil and gas related activities within their borders. The city of Munroe Falls, in Summit County, Ohio, applied its existing municipal code to require local drilling permits and hearing and certification requirements. Broadview Heights, in suburban Cleveland, and several other Ohio communities attempted to ban hydraulic fracturing by amending their local charters to prohibit it. These two local efforts differ in important ways. Munroe Falls attempted to control drilling by enforcing existing ordinances. Broadview Heights instituted a flat-out ban on oil and gas drilling not through ordinances, but by charter amendment. Ohio jurisdictions were testing the boundaries of their regulatory authority under a preempting statute and questionable home rule protection.

OHIO CONSTITUTION

The home rule provision of the Ohio Constitution gives municipalities "authority to exercise all powers of local self-government and to adopt and enforce within their limits such local police, sanitary and other similar regulations, as are not in conflict with general laws" (Ohio Const. art. XVIII, § 3). Cities can enact local self-government ordinances that conflict with state laws provided the state laws are not "general laws" (Gridley and Hardesty 2020, 4). So, when an Ohio municipality wants to control land use by setting spacing or set-back requirements for wells, it seems permissible under Ohio Constitution's home rule provision because set-back requirements do not feel "general" in nature. Traffic, noise, and other land use regulations are usually within the jurisdiction's home rule protection, but those bounds depend both on judicial interpretation and legislative acquiescence.

Through the 1990s, Ohio's government institutions supported strong home rule—leaving plenty of room for local control (Vaubel 1995; Brachman 2010). Areas of clear municipal authority include the structure and organization of local governments and procedures controlling local police powers (*State ex rel. Petit v. Wagner*). But courts upheld local actions to include, for example, animal control, fluoridation of water, and bans on "bait and switch" advertising—clearly beyond the workings of local government. Recently, the Ohio legislature began eroding municipal home rule—not only regarding oil and gas development, but in many areas of economic or political interest to the state. Ohio courts have restricted local authority in, for example, municipal residency requirements for public officials (Mulligan 2012, 366; *Am. Fin. Servs. Ass'n v. City of Cleveland*), municipal lending and housing regulations (Altier 2011; Ohio Rev. Code Ann. § 1332.22 (J) (2007); *Am. Fin. Servs. Ass'n v. City of Cleveland*), and cable television and video regulation (Saul 2009, 837; Ohio Rev. Code Ann. §§ 1332.21–1332.34 (2009)).

Ohio's regulatory balance has tipped away from local control and toward statewide control both through legislative restructuring of regulatory regimes and through courts' construction of the "general law" preemption doctrine (*City of Canton v. State*). The oil and gas industry maintains that Ohio's oil and gas law is a general law against which conflicting local rules would be void. Therefore, they argue local control of oil and gas related decision-making falls outside the protection of the Ohio Constitution's home rule provision.

OHIO LEGISLATION

The Ohio oil and gas law asserts state control over shale oil and gas regulation by assigning to the Ohio DNR "sole and exclusive authority" to regulate shale oil and gas activities. In addition, the legislation explicitly preempts local regulation of oil and gas related activities. Ohio's law also includes words like "matter of general statewide interest," "uniform statewide regulation," and "comprehensive plan" to indicate the legislature's intent to enact a general law which, when interpreted under the Ohio Constitution's home rule language, would preempt local regulation. Moreover, the legislature was specific and inclusive regarding the activities over which Ohio DNR has "sole and exclusive" authority. The statute now includes within state authority "all aspects of the locating, drilling, well stimulation, completing, and operating of oil and gas wells within this state, including site construction and restoration, permitting related to those activities, and the disposal of wastes." Presumably, the legislature expanded this list to encourage the Supreme Court of Ohio to interpret the statute as a general law preempting all local authority over shale oil and gas related activities.

THE SUPREME COURT OF OHIO

Unlike the courts in Pennsylvania and New York, Ohio's Supreme Court supported the state's efforts to control all aspects of oil and gas operations. As discussed, Ohio's Supreme Court has protected local regulatory authority in the past, but has increasingly found state laws to be "general laws" such as one preempting local regulation of guns and another preempting local residency requirements for police and fire employers (*Ohioans for Concealed Carry, Inc. v. City of Clyde; Missig v. City of Cleveland Civil. Serv. Comm'n.; Am. Fed'n & Mun. Emples. Local # 74 v. City of Warren*).

The Ohio Supreme Court addressed the question of whether a local ordinance that requires local permits and local mandates was preempted by the Ohio oil and gas statute. The Ohio Supreme Court's rule on the broader version of that question is that "a state statute takes precedence over a local ordinance when (1) the ordinance is in conflict with the statute, (2) the ordinance is an exercise of the police power, rather than of local self-government, and (3) the statute is a general law" (*City of Canton v. State*). The Ohio courts applied this rule to the specific question of whether the Ohio oil and gas law's preemption language is a general law and held that it is. Munroe Falls tried to enforce existing ordinances that would require oil and gas developers to acquire a local land use permit. Beck Energy, an oil and gas developer, had an Ohio DNR–issued drilling permit, but declined to seek a local Munroe Falls drilling permit, pay a fee, post a bond, wait a certain number of days to attend a public hearing regarding a land use variance, and meet other Munroe Falls requirements. The driller argued that all drilling activity was regulated only by the Ohio DNR, so Munroe Falls—a local government—could not add requirements without being in conflict with state law. Munroe Falls received an injunction to stop Beck Energy from drilling without complying with local law. To determine whether to uphold or reverse the injunction, the court of appeals evaluated the ordinances under the Ohio Supreme Court's home rule analysis (*City of Canton v. State*).

The Ohio legislature wrote that the regulation of oil and gas activities is a matter of general statewide interest that requires uniform statewide regulation, and Ohio's oil and gas law constitutes a comprehensive plan for such activities (Ohio Rev. Code Ann. 1509.02). With respect to Munroe Falls' ordinances requiring a conditional zoning certificate prior to drilling, the court of appeals found these to be exercises of police power, not self-government (City of Munroe Falls Ordinance §§ 1163.02, 1329.03, 1329.04, 1329.05, 1329.06; *State ex rel. Morrison v. Beck Energy Corp.*). In fact, all parties agreed that the city of Munroe Falls' rights of way ordinances were exercises of police power—designed to protect the public safety and general welfare, not self-government (*State ex rel. Morrison v. Beck Energy Corp.*). When an ordinance is an exercise of police power, as opposed to one of self-government, it does not fall within home rule protection and the court must decide whether the statute with which it conflicts is a "general law." If the statute is a general law, the statute trumps the local ordinance only if the ordinance conflicts with the statute (ibid.). The court of appeals cited a previous case indicating that the Ohio oil and gas statute "is unquestionably a general law" (*State ex rel. Morrison v. Beck Energy Corp.*; Ohio Rev. Code Ann. 1509.02; *Smith Family Tr. v. City of Hudson Bd. of Zoning & Bldg. Appeals*). It did not, therefore, pursue the general law analysis further, moving instead to the question of the ordinances' potential conflicts with state law (*State ex rel. Morrison; Smith Family Tr.*).

The court sought precedent to determine whether an ordinance conflicts with a statute. In one case, the court held that a municipal ordinance requiring increased fees and record keeping for waste facility operators could coexist with an apparently conflicting state statute regulating hazardous waste facilities (*Fondessy Enter. v. City of Oregon*). Although both concerned the monitoring of hazardous waste landfill facilities, they did not directly conflict with one another. The two government entities could exercise their respective police powers concurrently.

With respect to the Munroe Falls drilling ordinances, however, the court of appeals held that the state statute did *not* allow for additional local regulation (*State ex rel. Morrison*). Munroe Falls originally argued, and the trial court agreed, that the statute only limited local efforts to control permitting, location, and spacing of oil and gas wells (ibid.). Had Munroe Falls been faced with an earlier version of the statute this argument might have prevailed, but the legislature had expanded the statute's reach beyond the original permitting, location, and spacing. The court of appeals, therefore, disagreed with Munroe Falls and with the trial court, instead focusing on the statute's inclusive nature. The court of appeals was swayed by the language "comprehensive plan with respect to all aspects of the locating, drilling, well stimulation, completing, and operating of oil and gas wells within this state, including site construction and restoration, permitting related to those activities, and the disposal of wastes from those wells." The court of appeals found Munroe Falls' permit application fee and performance bond ordinances also to conflict with the state law. Because Munroe Falls' public hearing requirement was tied to its permit requirement, that too, was void. Munroe Falls' right of way and excavations ordinances, on the other hand, unlike those more closely related to drilling, did not conflict with the state law because of the statute's language specifically leaving the regulation of streets to local authorities.

The court of appeals removed the drilling injunction. It found that some of the Munroe Falls ordinances conflicted with Ohio's oil and gas law, which the court found, unlike the hazardous waste statute in *Fondessy*, was indeed a general law (*State ex rel. Morrison; Fondessy Enter. v. City of Oregon*). The Supreme Court of Ohio agreed.

PUTTING IT ALL TOGETHER IN OHIO

The legal environment Munroe Falls faced was difficult. Ohio legislation gave the Ohio DNR "sole and exclusive authority" to regulate most activities related to oil and gas development. The Ohio Constitution includes a home rule provision giving local jurisdictions some measure of local control, but the Ohio Supreme Court interpreted it increasingly narrowly, limiting its protections largely to decisions regarding the structure and function of local government.

After Munroe Falls obtained the drilling injunction, Beck Energy pursued the case and the Supreme Court of Ohio ultimately ruled in their favor. Munroe Falls was not allowed to enforce several of its municipal ordinances. The Supreme Court of Ohio had found Ohio's oil and gas law to be a general law delegating "sole and exclusive authority" over oil and gas regulation to Ohio DNR and preempting Munroe Falls' local authority to regulate oil and gas activities. This decision meant that Ohio local governments have little ability to control or influence drilling in their jurisdictions.

Pennsylvania

Pennsylvania is active in shale drilling, and its local jurisdictions, like their brethren in other states, want a voice in how and where it occurs. With Pennsylvania Act 13 (Pa. Cons. Stat. title 58 §§ 2301–3504 (2012)), much like Ohio's efforts, the state legislature amended the state's oil and gas laws to preempt local control of shale oil and gas production. Pennsylvania's highest court struck down this preemption, but not on home rule grounds. Like Munroe Falls in Ohio, Pennsylvania's Robinson Township sought to make local decisions regarding oil and gas permitting, but the state legal circumstances facing a local jurisdiction in Pennsylvania differ from those in Ohio.

PENNSYLVANIA CONSTITUTION

Like the Ohio Constitution, the Pennsylvania Constitution includes a home rule provision. Unlike Ohio, however, it also includes an Environmental Rights Amendment—a provision which proved valuable to local jurisdictions in Pennsylvania.

Pennsylvania's home rule language is broader than Ohio's. The Pennsylvania Constitution says that "a municipality which has a home rule charter may exercise any power to perform any function not denied by this Constitution, by its home rule charter or by the General Assembly at any time." Ohio limits home rule protection to local self-government and local police, sanitary regulation, and similar regulation not in conflict with general laws. Pennsylvania, however, allows any local controls that are not denied by the Pennsylvania Constitution or the General Assembly—which means by state legislation. Importantly, however, the Supreme Court of Pennsylvania did not rely on this broader home rule provision in its dispositive case on local control. Instead, it relied on the Environmental Rights Amendment (*Robinson Twp. v. Commonwealth of Pennsylvania*, 83 A.3d 901).

In addition to its home rule provision, Pennsylvania's constitution includes language for which there is no comparable provision in Ohio's constitution. Pennsylvania's constitution includes an Environmental Rights Amendment stating:

> The people have a right to clean air, pure water, and to the preservation of the natural, scenic, historic and esthetic values of the environment. Pennsylvania's public natural resources are the common property of all the people, including generations yet to come. As trustee of these resources, the Commonwealth shall conserve and maintain them for the benefit of all the people. (Pa. Const. art. I, §27)

Pennsylvania citizens adopted the amendment by referendum in 1971 when public awareness of environmental pollution from industrialization, mining activities, and other damaging industries also led to many of the federal environmental laws (Pa. Code title 25 §127.81–83; Dernbach, May, and Kristl 2015, 1171). The Pennsylvania Supreme Court relied on the Environmental Rights Amendment to give local jurisdictions control over zoning decisions applicable to oil and gas activities (*Robinson Twp.*, 83 A.3d 901).

PENNSYLVANIA LEGISLATION

Pennsylvania faced issues similar to Ohio's concerning the intersection of state oil and gas laws and local efforts to control activities within municipal boundaries. In Pennsylvania, the legislature enacted Act 13, in part an attempt to preempt local regulation. The Pennsylvania oil and gas law was originally enacted in 1961 and did not address horizontal drilling or hydraulic fracturing technologies that were quickly arriving to the Marcellus Shale play. Because of this omission, these technologies were regulated in a "patchwork of regulatory responses by the state as well as local governments" (Dernbach, May, and Kristl 2015, 1171). As it modernized its statute, like the Ohio legislature, Pennsylvania's legislature added language to preempt local regulation of oil and gas operations.

Act 13's section 3303 said with respect to oil and gas development that the legislature intends to *"occupy the entire field of regulation, to the exclusion of all local ordinances."* It also said that "this section, *preempts and supersedes the local regulation of oil and gas operations"* (emphasis added). In section 3304, the act directed local jurisdictions to "allow for the reasonable development of oil and gas resources" (Pa. Cons. Stat. title 58 §3304(a) (2012), invalidated by *Robinson Twp.*, 52 A.3d 463). Act 13 required localities to allow "well and pipeline location assessment operations" (Pa. Cons. Stat. title 58 § 3304(b)(1)), prohibited them from imposing requirements more stringent than the state's with respect to drilling construction and operations (ibid. §3304(b)(2)–(3)), and required them to complete any local review of drilling proposals within time limits specified by statute (ibid. §3304(b)(5)).

Robinson Township, in Allegheny County near Pittsburgh, and several other Pennsylvania townships sued Pennsylvania, challenging Act 13, among other things, as inconsistent with the Pennsylvania Constitution's Environmental Rights Amendment. They argued that the amendment allowed, or even required, government entities to make decisions to protect citizens' environmental rights.

THE SUPREME COURT OF PENNSYLVANIA

Robinson Township argued that Act 13 negated its police power based authority to control, among other things, where wells could be safely located. As written, Act 13 would force changes in zoning and traffic laws, thereby usurping Robinson Township's home rule powers and its responsibility to its citizens under the Environmental Rights Amendment (ibid.).

In July 2012, the Pennsylvania Commonwealth Court, an intermediate court acting here as the trial court, struck down section 3304 as unconstitutional because it would have required incompatible land uses to persist within local zoning districts (*Robinson Twp.*, 52 A.3d 463). The Commonwealth Court considered the preemption language in section 3303 but upheld it despite Environmental Rights Amendment challenges (ibid.). Earlier Pennsylvania court decisions had held that the Environmental Rights Amendment applies only when specifically invoked by the legislature within the legislation itself (Dernbach, May, and Kristl 2015). The Pennsylvania legislature had not invoked the Environmental Rights Amendment in Act 13.

The Pennsylvania Supreme Court upheld the Commonwealth Court's voiding of the 3304 provisions which had unconstitutionally imposed upon local control by requiring that all local ordinances provide for the "reasonable development of oil and gas resources." The court returned to local governments the authority to regulate hydraulic fracturing as an industrial use under their land use and zoning ordinances. It also declared unconstitutional Act 13's requirement in Section 3303 that all municipal ordinances regulating oil and gas operations be uniform and its mandate that certain oil and gas related activities be allowed in every zoning district. These Act 13 requirements, had they been allowed to continue, would have eliminated local governments' ability to consider oil and gas operations in their zoning plans.

One essential issue for the Pennsylvania Supreme Court in the *Robinson Township* case was whether the Environmental Rights Amendment gives rights to citizens and responsibilities to municipalities or is merely a guide for the legislature. The Commonwealth argued that the Environmental Rights Amendment is merely a guide to the legislature as it decides what is best for the citizens, the natural resources, and the environment of Pennsylvania. The municipalities argued that the Environmental Rights Amendment confers rights upon citizens and responsibilities upon its municipalities. If the amendment gave municipalities responsibilities, they would have to be able to exercise them by creating rules or standards in their communities.

On this point, the Pennsylvania Supreme Court's plurality sided with the municipalities, partly because the Environmental Rights Amendment is located in the state constitution's equivalent of the Bill of Rights (Dernbach, May, and Kristl 2015, 1171). The Supreme Court of Pennsylvania considered the plain language of the amendment which says, "the people have a right to clean air, pure water, and to the preservation of the natural, scenic, historic and esthetic values of the environment." This appeared to confer a right on the people—to clean air, pure water, and the preservation of certain environmental values. Inherent in that sentence, the court also found a "limitation on the state's power to act contrary to this right" (Dernbach, May, and Kristl 2015, 1171).

The amendment continues: "Pennsylvania's public natural resources are the common property of all the people, including generations yet to come. As trustee of these resources, the Commonwealth shall conserve and maintain them for the benefit of all the people" (*Robinson Twp.*, 83 A.3d 901; Pa. Const. art. I, §27). The court said, "the plain meaning of the terms conserve and maintain implicates a duty to prevent and remedy the degradation, diminution, or depletion of our public natural resources" (*Robinson Twp.*, 83 A.3d 901). The court also found a duty "to act affirmatively to protect the environment, via legislative action" (*Robinson Twp.*, 83 A.3d 901). So, rather than merely providing guidance for the legislature, the Environmental Rights Amendment imposed an affirmative duty upon the legislature and also upon municipalities to protect citizens' environmental rights.

PUTTING IT ALL TOGETHER IN PENNSYLVANIA

In July 2012, in *Robinson Township v. Commonwealth of Pennsylvania*, the Commonwealth Court (an intermediate court) held that the provisions of Act 13 that would override local zoning and environmental laws were unconstitutional, null, and void (*Robinson Twp.*, 52 A.3d 463). The Pennsylvania court found that those provisions violated the Pennsylvania Constitution's Environmental Rights Amendment. Despite the Pennsylvania legislature's efforts to give the state complete control over oil and gas decision-making, the Pennsylvania Constitution's Environmental Rights Amendment preserved the ability of localities to use their zoning and land use powers in the area of shale oil and gas development. By contrast, the Ohio Constitution, although similar to Pennsylvania's in the home rule area, does not include a provision on environmental rights, so the Pennsylvania decision was neither helpful nor influential for Ohio's city of Munroe Falls.

New York

Like Ohio and Pennsylvania, New York has a home rule provision in its constitution and a statute stating that New York law supersedes local ordinances regulating oil and gas. Still, even more abundantly and enthusiastically than their counterparts in Ohio and Pennsylvania, local governments in New York enacted ordinances purporting to exert control over oil and gas activities. Many of the local actions in New York went far beyond attempts to control or influence oil and gas activities, often including language banning drilling or the use of hydraulic fracturing. For example, the Town of Dryden amended its zoning rules to prohibit hydraulic fracturing (*Wallach v. Town of Dryden*). The Town of Middlefield also enacted a ban (*Cooperstown Holstein Corp. v. Town of Middlefield*). Both were sued—Dryden by an energy company that had acquired oil and gas leases prior to the zoning amendment, and Middlefield by a dairy farm that had leased its land to a drilling company. In both cases, the drilling company argued that the towns had no authority to enact drilling bans because New York law preempted those actions.

The New York Court of Appeals upheld a lower court decision supporting the local ordinances on home rule grounds. The lower court found, and the high court agreed, that the state statute did not eliminate the towns' authority to ban hydraulic fracturing using their zoning authority. New York's Oil, Gas, and Solution Mining Law, which attempted to preempt local control, could not supersede the New York Constitution's home rule provision, despite its intent to suppress local regulation of land use (*Wallach v. Town of Dryden*). The court seemed to say that New York values local control of land use decisions, as evidenced by its constitution's home rule provision. The legislature's attempt to emasculate that provision by including suppression language in the oil and gas law was not effective. The state statute could not trump the state constitution.

More than one hundred and fifty New York jurisdictions passed bans on shale oil and gas drilling, or on the use of the hydraulic fracturing technology (Cecot 2018). They enacted these bans knowing that state law purported to preempt them. New York's highest court would ultimately uphold their bans on state home rule and environmental grounds resident in New York's constitution. But in the meantime, New York's governor had placed a temporary administrative ban on hydraulic fracturing pending an environmental review by the New York

Department of Environmental Conservation (New York DEC). Communities were concerned that once the study was completed and the moratorium lifted, it would be too late to take local action. They wanted bans in place prior to the end of the moratorium—the date of which was unknown.

NEW YORK CONSTITUTION

Like the constitutions of Ohio and Pennsylvania, New York's constitution has a home rule provision. Article IX of the New York Constitution says, "every local government shall have power to adopt and amend local laws not inconsistent with the provisions of this constitution or any general law ... except to the extent that the legislature shall restrict the adoption of such a local law" (N.Y. Const. art. IX, §2(c)(ii)).

NEW YORK LEGISLATION

The primary New York legislation influencing local jurisdictions' ability to control shale oil and gas development includes home rule implementing statutes, an oil and gas law, and a state environmental review act. The New York legislature enacted the Municipal Home Rule Law, a state statute, to implement the state constitution's home rule provision. This empowers local governments to pass laws both for the "protection and enhancement of [their] physical and visual environment" (N.Y. Mun. Home Rule Law, §10(1)(ii)(a)(11) (McKinney 2014)) and for the "government, protection, order, conduct, safety, health and well-being of persons or property therein" (N.Y. Mun. Home Rule Law, §10(1)(ii)(a) (12) (McKinney 2014)). In addition, the New York legislature enacted the Town Law authorizing towns to enact zoning laws to foster "the health, safety, morals, or the general welfare of the community" (N.Y. Town Law §261 (McKinney 2014)) and the Statute of Local Governments, which granted towns the power to "adopt, amend, and repeal zoning regulations" (N.Y. Stat. Loc. Govs. §10(6) (McKinney 2014)).

Article 23, New York's Oil, Gas and Solution Mining Law (OGSML) (N.Y. Envtl. Conserv. Law, §23 (McKinney 2014)) gives the New York Department of Environmental Conservation responsibility for shale oil and gas regulation (ibid.). The legislature's policy statement, associated with Act 23, says, "this title shall supersede all other state and local laws relating to the extractive mining industry" (N.Y. Envtl. Conserv. Law, §23–2703 (McKinney 2014)). But the clause includes important caveats not found in Ohio or Pennsylvania law. The statute does not prohibit local governments from "enacting or enforcing local laws or ordinances of general applicability" unless they regulate mining that the state has already regulated (ibid., 2(a)). The statute contains similar subsections allowing for local zoning regulations and allowing localities to regulate in circumstances for which the state does not require permitting (ibid., 2(b)(c)). The tension in New York is in the degree to which the policy document and supersession clauses conflict with home rule language elsewhere in New York's legal and political traditions.

Unlike most states, New York has the State Environmental Quality Review Act (SEQRA). Known as a "little NEPA" law, it is a state version of the National Environmental Policy Act, or NEPA. NEPA requires federal agencies to consider potential environmental effects prior to making decisions that could significantly affect the human environment. New York's little NEPA goes beyond the federal law. The SEQRA requires New York agencies not only to consider the environmental impact of their decisions but also to act in the best interest of the environment according to an environmental review. The New York DEC did this required review in 1992, called a General Environmental Impact Statement (GEIS), prior to its adoption of its Oil, Gas and Solution Mining Regulatory Program.

Years later, the oil and gas industry changed its production process to include the use of high-volume hydraulic fracturing. This led to a renewed focus on oil and gas production regulation and concern about the environmental and health impacts of the practice. New York agencies began to reassess the state's regulatory program. To do that required agency compliance with the SEQRA and accompanying environmental review.

In 2008, the New York legislature passed a moratorium on high-volume hydraulic fracturing (HVHF). When it expired in 2010, the legislature proposed a bill to limit drilling (Sadasivam 2014). The then governor Paterson vetoed it, but issued an executive order establishing another six-month drilling moratorium. The next governor, Andrew Cuomo, continued this executive moratorium. Because an administrative ban on HVHF would change executive branch policy it was effectively an additional agency action requiring environmental review under the SEQRA, beyond the 1992 GEIS. The New York DEC needed to produce a Supplemental Generic Environmental Impact Statement (SGEIS) regarding high-volume hydraulic fracturing.

To discharge this duty, the New York DEC asked the New York State Department of Health (New York DOH) to review the DEC findings regarding whether health impact mitigation measures contained in a Draft SGEIS were adequate to protect public health from the risks of high-volume hydraulic fracturing. In December 2014, the New York DOH published "A Public Health Review of High-Volume Hydraulic Fracturing for Shale Gas Development" (New York State Department of Public Health 2014). In it, the DOH advised the DEC that several potential adverse environmental impacts could result from high-volume hydraulic fracturing that may be associated with adverse public health outcomes.

These impacts include (1) air impacts that could affect respiratory health due to increased levels of particulate matter, diesel exhaust, or volatile organic chemicals; (2) climate change impacts due to methane and other volatile organic chemical releases to the atmosphere; (3) drinking water impacts from underground migration of methane or fracturing fluid chemicals associated with faulty well construction or seismic activity; (4) surface spills potentially resulting in soil, groundwater, and surface water contamination; (5) surface water contamination resulting from inadequate wastewater treatment; (6) earthquakes and creation of fissures induced during the hydraulic fracturing stage; and (7) community character impacts such as increased vehicle traffic, road damage, noise, odor complaints, and increased local demand for housing and medical care. New York DOH concluded, therefore, that "until the science provides sufficient information to determine the level of risk to public health from HVHF to all New Yorkers and whether the risks can be adequately managed . . . HVHF should not proceed in New York State" (SGEIS). In other words, New York DEC's mitigation measures contained in the Draft SGEIS were not sufficient to allow hydraulic fracturing to continue. Based on the study's findings of significant public health risks tied to air pollution and water contamination, as well as insufficient evidence to affirm the technology's safety, Governor Cuomo and the New York DEC instituted the statewide ban on hydraulic fracturing (Kaplan 2014).

The final SGEIS, which included the New York DOH's health effects study, was required by the SEQRA and resulted in the banning of high-volume hydraulic fracking in New York. Although not tied directly to the regulatory authority of local governments to act on oil and gas related issues within their boundaries, the SEQRA provides them an additional avenue for intervention, involvement, and influence not available in other Marcellus states. In New York, although the Court of Appeals supported local efforts to control or ban oil and gas related activities, the SEQRA provided opportunities for them to exert influence through a statewide public process, potentially obviating their own local actions.

NEW YORK COURT OF APPEALS

Separate from the SEQRA process that led to a New York State ban on hydraulic fracturing, the highest court in New York, the New York Court of Appeals, supported local governments' efforts to retain control regarding oil and gas activities. It said in *Wallach v. Town of Dryden*, "the [New York] Legislature has recognized that local regulation of land use is 'among the most significant powers and duties granted . . . to a town government.'" The court also wrote that local control of zoning is "one of the core powers of local governance." The court said, "a town may not enact ordinances that conflict with the State Constitution or any general law." It noted that "under the preemption doctrine, a local law promulgated under a municipality's home rule authority must yield to an inconsistent state law." The court held the legislature's suppression of local zoning control to be sufficiently important that it allowed it only with a "clear expression of legislative intent to preempt local control over land use."

The court, therefore, considered whether the legislature evidenced sufficiently clear intent to preempt local control over land use decisions. It started with the statutory language: "The provisions of [the OGSML] shall supersede all local laws or ordinances relating to the regulation of the oil, gas and solution mining industries; but shall not supersede local government jurisdiction over local roads or the rights of local governments under the real property tax law" (*Wallach v. Town of Dryden*). Unsurprisingly, the towns' opponents argued that the legislature's suppression language should "be interpreted broadly to reach zoning laws that restrict, or as presented here, prohibit oil and gas activities, including hydro-fracking, within municipal boundaries" (ibid.). Like the Ohio courts, New York has a three-part test for determining whether a suppression clause is sufficiently clear to preempt a local ordinance. The New York test looks to "(1) the plain language of the supersession clause; (2) the statutory scheme as a whole; and (3) the relevant legislative history" (ibid.).

In *Matter of Frew Run*, the town created a zoning district that would prohibit sand and gravel production (*Matter of Frew Run Gravel Prods. v. Town of Carroll*). A disappointed sand and gravel company pointed to New York's Mined Land Reclamation Law (MLRL) and argued that state law superseded the town's zoning ordinance. The MLRL said, "this title shall supersede all other state and *local laws relating to the extractive mining industry*; provided, however, that nothing in this title shall be construed to prevent any local government from enacting local zoning ordinances or other local laws which impose stricter mined land reclamation standards or requirements than those found herein" (*Wallach v. Town of Dryden*; emphasis added). The court found that "local laws relating to the extractive mining industry" did not include the town's zoning provisions. It held that zoning laws are not laws related to the extractive mining industry. Instead, they related to land use decisions within the town. The court found this to be "an entirely different subject matter and purpose." The court distinguished between local rules that control operations and processes concerning mining, which would be superseded by the state law, and those that pertain to land use questions, which instead are a matter for local discretion.

The court found the circumstances for the towns of Dryden and Middlesex to be substantially similar to those in *Matter of Frew*. Still, it applied the three-part test. First, it looked at the plain language of the statute. The NY oil and gas law preempts local laws "relating to the regulation of the oil, gas and solution mining industries" whereas the state statute in *Matter of Frew* purported to preempt local laws "relating to the extractive mining industry." As it did in *Matter of Frew*, the court found that local zoning ordinances are not local laws relating to a specific industry, here the regulation of the oil, gas, and solution mining industries. Instead, as in *Matter of Frew*, they relate to how land may be used within the town borders—an issue of local concern.

PUTTING IT ALL TOGETHER IN NEW YORK

The New York Court of Appeals paved the way for local governments in New York to use local zoning ordinances to control drilling within their borders. The court did this by interpreting local ordinances, for example, banning hydraulic fracturing and on location and spacing of wells, to be local land use controls and therefore not in conflict with the state oil and gas law that might otherwise supersede them.

The legal mechanism through which New York State's ban on HVHF took effect, however, is unique among the Marcellus-region states. Most often— as was the case in Ohio—drilling laws have taken the form of statewide laws, passed through the legislative body, or regulation by the assigned state agencies. In New York and Pennsylvania, thanks to state supreme court interpretations of their constitutions, local ordinances play a role. But New York was also able to take a different approach. Governor Cuomo's administration, specifically the New York DEC, issued an official finding prohibiting the use of high-volume hydraulic fracturing (Associated Press 2015; N.Y. Envtl. Conserv. Law, § 23; New York State Department of Environmental Conservation 2015). The prohibition is legally binding. New York's SEQRA, which requires agencies to conduct and then act upon the results of environmental review prior to decision-making made this route possible. It is a route unavailable in the other Marcellus-region states because they lack a "little NEPA" among their state laws.

Local Governments Going Forward

Little NEPAs

About fifteen states have their own versions of the National Environmental Policy Act (NEPA) like New York's SEQRA, which laid the groundwork for the state's administrative ban on high-volume hydraulic fracturing. Ohio and Pennsylvania are not among them. Ohio and Pennsylvania, therefore, do not have the same legal tools available to them to control oil and gas drilling activities. Although this does not bear directly on the regulatory authority of local governments, New

York's SEQRA provided an opportunity in law that is unavailable in most jurisdictions. If Ohio and other states had "little NEPA" laws they would require state
agencies to study and consider the environmental impact of their decisions—and
perhaps even to act in accordance with what they learned.

Constitutional Environmental Rights

Ohio, like many other states, lacks an environmental rights provision in its constitution similar to that which allowed Pennsylvania to return regulatory authority
over zoning for oil and gas activities to local governments. The Ohio Constitution declares environmental conservation to be a proper public purpose of state
and local governmental entities (Ohio Const. art. VIII, § 2), but does not confer
rights on the citizens or responsibility on state or local government. Environmental scholars have written about state and local environmental constitutionalism
(Dernbach, May, and Kristl 2015, 1171; Craig 2004) noting that Pennsylvania is
not the only state constitution to offer environmental rights and that the *Robinson Township* decision may inspire other courts to confer environmental rights
on citizens and responsibilities on state and local governments via constitutional
interpretation (Dernbach, May, and Kristl 2015; Craig 2004).

Community Bills of Rights

Facing obstacles to local control in their state legislatures and courts, environmental organizations have been working with local jurisdictions to amend their
charters to include Community Bills of Rights (CBRs). While also asserting other
rights on behalf of local citizens, CBRs often prohibit oil and gas related activities.
These efforts have not been effective in terms of changing the role of local jurisdictions or the citizens they represent because they are likely unenforceable in states
that preempt local control. That said, states without environmental rights provisions in their constitutions, like Ohio, would benefit from environmental rights
protection of some sort. Perhaps starting locally, as the CBR movement has done,
will help bring light to the issue statewide. The ultimate goal would be to have an
environmental rights provision in the Ohio Constitution, like that in Pennsylvania.

Using Traditional Land Use Control Authority

One of the Ohio Supreme Court opinions in the Munroe Falls case, a concur-
rence by Justice Terrance O'Donnell, noted that rather than using local ordinances which a court could find to conflict with state law, local jurisdictions
should use traditional zoning authority to protect their communities. They could

use noise ordinances and residential and commercial zoning restrictions. Justice O'Donnell suggested that if local jurisdictions were to do this—use traditional zoning, nondiscriminatorily created and nondiscriminatorily applied to oil and gas activities—this might withstand state preemption of local authority.

Banding Together as a Region

Another method of potential local control or influence lies in banding together as regions. Gates Mills, Ohio, for example, created a regional commission to work with local political leaders throughout northeast Ohio to determine how they might adapt their zoning codes in compliance with Justice O'Donnell's opinion. Their goal is to adapt zoning codes to influence drilling in whatever directions suit their communities, within the boundaries of the existing legal system. Of course, they cannot regulate drilling activities directly because that would violate the Ohio Supreme Court's interpretation of the Ohio legislature's delegation of "sole and exclusive" authority to the state agency, as well as its preemption of local control. Use of traditional zoning authority must not be discriminatory, including in application to oil and gas activities.

The Bottom Line

This chapter showed how state legal environments shape and constrain prospects for local governments in Ohio, Pennsylvania, and New York to control drilling. It describes some of the complex factors interwoven in state governance through the language and judicial interpretation of state constitutions and state statutes. In each state, legal institutions—which differed substantially—shaped the contexts within which local governments must operate. State constitutions, legislation, and judicial interpretations are critical to the regulatory actions local governments may take regarding unconventional well drilling in their communities. Local regulatory efforts can be supported, or largely derailed, depending on their state legal environment.

References

Act 13. See Pa. Cons. Stat. title 58 §§ 2301–3504 (2012).

Altier, Brett. 2011. "Municipal Predatory Lending Regulation in Ohio: The Disproportionate Impact of Preemption in Ohio's Cities." *Cleveland State Law Review* 59 (1): 125–60.

Am. Fed'n & Mun. Emples. Local # 74 v. City of Warren, 895 N.E.2d 238 (Ohio Ct. App. 2008).

Am. Fin. Servs. Ass'n v. City of Cleveland, 858 N.E.2d 776 (Ohio 2006).

Angarola, Jonathon. 2015. "Ohio's Home-Rule Amendment: Why Ohio's General Assembly Creating Regional Governments Would Combat the Regional Race to the Bottom under Current Home-Rule Principles." *Cleveland State Law Review* 63 (4): 865–99.

Associated Press. 2015. "New York Releases Final Environmental Review of Fracking." *Auburn Citizen*, May 14. http://auburnpub.com/news/local/state-and-regional/new-york-releases-final-environmental-review-of-fracking/article_42bcccdc-f9ef-11e4-9d6e-6b231511973f.html.

Brachman, Lavea. 2010. "Legislating Sustainable Design: The Challenge of Local Control and Political Will." *Environmental Law Reporter* 40 (8): 10740–42.

Cecot, Caroline. 2018. "No Fracking Way: An Empirical Investigation of Local Shale Development Bans in New York." *Environmental Law* 48:761–95.

City of Canton v. State, 766 N.E.2d 963 (Ohio 2002).

City of Munroe Falls Ordinance § 1163.02 (2017).

City of Munroe Falls Ordinance § 1329.03 (2017).

City of Munroe Falls Ordinance § 1329.04 (2017).

City of Munroe Falls Ordinance § 1329.05 (2017).

City of Munroe Falls Ordinance § 1329.06 (2017).

Cooperstown Holstein Corp. v. Town of Middlefield, 964 N.Y.S.2d 431 (N.Y. 2013).

Craig, Robin Kundis. 2004. "Should There Be a Constitutional Right to a Clean/Healthy Environment?" *Environmental Law Reporter* 34 (12): 11013–24.

Dernbach, John C. 1999a. "Taking the Pennsylvania Constitution Seriously When It Protects the Environment: Part I—An Interpretative Framework for Article I, Section 27." *Dickinson Law Review* 103 (4): 693–34.

Dernbach, John C. 1999b. "Taking the Pennsylvania Constitution Seriously When It Protects the Environment: Part II—Environmental Rights and Public Trust." *Dickinson Law Review* 104 (1): 97–164.

Dernbach, John C., James May, and Kenneth Kristl. 2015. "Robinson Township v. Commonwealth of Pennsylvania: Examinations and Implications." *Rutgers University Law Review* 67 (5): 1169–96.

Fesler, Mayo. 1916. "The Progress of Municipal Home Rule in Ohio." *National Municipal Review* 5 (2): 242–51.

Fondessy Enter. v. City of Oregon, 492 N.E.2d 797, 799–800 (Ohio 1986).

Gridley, Wendy H., and Amber Hardesty. 2020. "Municipal Home Rule: Members Only; An Informational Brief Prepared for Members of the Ohio General Assembly by the Legislative Service Commission Staff." 133 (5): 1–13. https://www.lsc.ohio.gov/documents/reference/current/membersonlybriefs/133Municipal%20Home%20Rule.pdf.

Kaplan, Thomas. 2014. "Citing Health Risks, Cuomo Bans Fracking in New York State." *New York Times*, December 17. http://www.nytimes.com/2014/12/18/nyregion/cuomo-to-ban-fracking-in-new-york-state-citing-health-risks.html?_r=0.

Matter of Frew Run Gravel Prods. v. Town of Carroll, 71 N.Y.2d 126 (N.Y. 1987).

Missig v. City of Cleveland Civil Serv. Comm'n, 915 N.E.2d 642 (Ohio 2009).

Mulligan, Joe. 2012. "Not in Your Backyard: Ohio's Prohibition on Residency Requirements for Police Officers, Firefighters, and Other Municipal Employees." *University of Dayton Law Review* 37 (3): 351–79.

New York State Department of Environmental Conservation. 2015. "Final Supplemental Generic Environmental Impact Statement on the Oil, Gas and Solution Mining Regulatory Program." May 2015. http://www.dec.ny.gov/docs/materials_minerals_pdf/fsgeis2015.pdf.

New York State Department of Health. 2014. "A Public Health Review of High Volume
    Hydraulic Fracturing for Shale Gas Development." December 2014. https://dela
    warehighlands.org/wp-content/uploads/NYS_DOH_high_volume_hydraulic_
    fracturing.pdf.
N.Y. Const. art. IX, § 2(c)(ii).
N.Y. Envtl. Conserv. Law § 23 (McKinney 2014).
N.Y. Envtl. Conserv. Law § 23–2703 (McKinney 2014).
N.Y. Mun. Home Rule Law § 10(1)(ii)(a)(11) (McKinney 2014).
N.Y. Mun. Home Rule Law § 10(1)(ii)(a)(12) (McKinney 2014).
N.Y. Stat. Loc. Govs. § 10(6) (McKinney 2014).
N.Y. Town Law § 261 (McKinney 2014).
Ohioans for Concealed Carry, Inc. v. City of Clyde, 896 N.E.2d 967 (Ohio 2008).
Ohio Const. art. VIII, § 2.
Ohio Const. art. XVIII, § 3.
Ohio Rev. Code Ann. § 1332.22 (J) (West 2007).
Ohio Rev. Code Ann. §§ 1332.21–1332.34 (West 2009).
Ohio Rev. Code Ann. § 1509.02 (West 1964).
Pa. Code title 25 § 127.81–83 (2012).
Pa. Cons. Stat. title 58 §§ 2301–3504 (2012) (Act 13).
Pa. Cons. Stat. title 58 § 3304(a) (2012).
Pa. Cons. Stat. title 58 § 3304(b)(1) (2012).
Pa. Cons. Stat. title 58 § 3304(b)(2)–(3) (2012).
Pa. Cons. Stat. title 58 § 3304(b)(5) (2012).
Pa. Const. art. I, § 27.
Richardson, Nathan, Madeline Gottlieb, Alan Krupnick, and Hannah Wiseman. 2013.
    "The State of State Shale Gas Regulation." Resources for the Future. https://
    media.rff.org/documents/RFF-Rpt-StateofStateRegs_Report.pdf.
Robinson Twp. v. Commonwealth of Pennsylvania, 52 A.3d 463 (Pa. Commw. Ct. 2012).
Robinson Twp. v. Commonwealth of Pennsylvania, 83 A.3d 901 (Pa. 2013).
Sadasivam, Naveena. 2014. "New York State of Fracking: A ProPublica Explainer."
    ProPublica, July 22. http://www.propublica.org/article/new-york-state-of-
    fracking-a-propublica-explainer.
Saul, Gregory M. 2009. "Constitutional Issues under Ohio's New Regulatory Frame-
    work for Video Service Providers." *Capital University Law Review* 37 (3):
    819–57.
Smith Family Tr. v. City of Hudson Bd. of Zoning & Bldg. Appeals, No. C.A. 24471,
    2009 Ohio App. LEXIS 2251 (Ohio Ct. App. June 3, 2009).
State ex rel. Morrison v. Beck Energy Corp., 37N.E.3 d 128 (Ohio 2015).
State ex rel. Petit v. Wagner, 164 N.E.2d 574, 576 (Ohio 1960).
Vaubel, George D. 1995. "Municipal Home Rule in Ohio (1976–1995)." *Ohio Northern
    University Law Review* 22 (1): 143–248.
Vill. of W. Jefferson v. Robinson, 205 N.E.2d 382 (Ohio 1965).
Walker, Harvey. 1948. "Municipal Government in Ohio before 1912." *Ohio State Law
    Journal* 9 (1): 1–17.
Wallach v. Town of Dryden, 16 N.E.3d 1188 (N.Y. 2014).
Wiseman, Hannah J. 2014. "The Capacity of States to Govern Shale Gas Development
    Risks." *Environmental Science and Technology* 48 (15): 8376–87.

HOW THE LEGAL FRAMEWORK OF FRACKING IN APPALACHIA DISSERVES THE POOR

Ann M. Eisenberg

About a decade ago, the attention of many commentators in the energy sphere turned to new technological developments that made bodies of subterranean natural gas more financially accessible to developers. The Marcellus and Utica gas deposits drew particular attention because of the volume of gas they held under layers of shale rock. The Marcellus body held an estimated 500 trillion cubic feet of natural gas, while the Utica shale deposit held an estimated 5 to 15 trillion cubic feet—enough to meet the nation's gas demand for decades (Eisenberg 2015). Developers would be able to extract this gas through the novel method of high-volume hydraulic fracturing, also known as "fracking." Fracking involves blasting the shale rock with a mixture of chemicals in order to crack it and release the gas, then pumping the gas back out.

Stretching from southern New York through Pennsylvania and into Ohio, West Virginia, and Maryland, the newly accessible gas in the Marcellus and Utica shale bodies represented a potential energy revolution in the minds of many. Depending on how greenhouse gas emissions are measured, burning natural gas may result in lower emissions than coal. While it is still a carbon-emitting fossil fuel, fracking's proponents argued that it could be a "bridge fuel" to a more deeply decarbonized future. Of course, many also stood to make substantial profits by taking advantage of the "fracking boom." Fracking advocates argued that these profits would flow to struggling rural economies in the form of new jobs in the industry and hearty royalty payments for the owners of mineral rights (Eisenberg 2015; Fish 2011).

Skeptics, however, raised concerns about the novel technology. Little information was available about the technical aspects of fracking, including potential environmental impacts. Additionally, because the technology was so new, legal regimes did not seem ready to manage this new era in natural resource extraction. While it was clear there was money to be made, many questioned whether the potential environmental ramifications would be worth the potential economic benefits (Eisenberg 2015).

This chapter examines one particular way in which the so-called fracking revolution has not been so revolutionary. Specifically, it illustrates first that the public discourse on fracking in the Marcellus and Utica shale region neglected a critical circumstance: these shale deposits are located in Appalachia. Their location in Appalachia is significant because Appalachia already has a sordid history with natural resource extraction. The public discourse on fracking has mostly neglected the fact that Appalachia is already "coal country" (Bell and York 2010). Yet, serving as the nation's energy powerhouse has not served Appalachia well, financially or otherwise (Bell and York 2010; Caudill 1963). The oversight of Appalachia's historical role as a "sacrifice zone"—providing natural resources at great cost for the benefit of the rest of the country (Eisenberg 2015)—casts doubt on arguments in favor of a liberal regulatory fracking regime, and perhaps casts doubt on the wisdom of pursuing fracking altogether. This history especially undermines the arguments insisting that natural resource extraction can bring economic development benefits to Appalachia, when the experience to date has been just the opposite.

This chapter further illustrates that certain aspects of the legal regimes governing fracking in three Appalachian states—Ohio, Pennsylvania, and West Virginia—also cast doubt on the argument that shale gas development would bring renewed prosperity to struggling Appalachians. As we saw in chapter 2, New York banned fracking, and this chapter adds West Virginia to the case analysis above. While many landowners have made substantial profits from the shale gas underlying their land, the benefits have been unevenly distributed (Clough and Bell 2016). These three states' legal regimes help explain this uneven distribution in part. In certain instances where the law might serve to protect those who may be financially or physically vulnerable regarding shale gas development, the law has failed to intervene. Conversely, the law has intervened in ways that may further disadvantage the vulnerable. In some ways, the law has just begun to catch up, even though the boom now appears to be headed toward a "bust" phase.

After a brief review of Appalachia's tragic historical relationship with natural resource extraction, this chapter examines two ways in which Ohio's, Pennsylvania's, and West Virginia's respective legal regimes have served to benefit corporate interests instead of the less privileged residents of this region. Specifically, the chapter reviews two central failures to empower landowners and communities to be better protected. First, these states' laws have failed to adequately regulate the contract-seeking professionals known as "landmen" or "land agents," resulting in systemic unfairness in negotiations between companies and landowners. Second, these states' legal regimes have failed to adequately ensure environmental justice and prevent environmental injustice.

Environmental injustice characterizes the Appalachian story. The 1980s movement that originated the demand for environmental justice, or EJ, was initially concerned with environmental racism, or the fact that minority communities bear a disproportionate burden of environmental hazards (Eisenberg 2015). EJ advocates today call for environmental hazards to be more equitably distributed across racial lines and other categories, such as class, disability, and veteran status. Appalachia's environmental and human sacrifices to benefit the coal industry and the rest of the country are a classic story of a vulnerable population bearing more than its fair share of environmental costs. The fracking boom and the absence of robust regulations that explicitly recognize EJ raise the prospect of similar concerns. Many rural residents are "land-rich, cash-poor," so even someone owning a significant amount of property could be rendered vulnerable to economic and physical exploitation (Eisenberg 2015). Those who do not own land still experience impacts from the dramatic changes brought to communities by shale gas development (Clough and Bell 2016).

Background on Appalachia and Natural Resource Extraction

From the early days of the nation's history, speculators and business owners on the east coast recognized that the central Appalachian Mountains contained a wealth of natural resources, the exploitation of which could yield substantial profits. Timber and coal in particular drew developers' attention. Standing in between the profits and the developers, however, were the Scotch Irish settlers who had moved into the mountains in the seventeenth and eighteenth centuries (Caudill 1963; Stoll 2017).

The "mountaineers" tended to stay aloof from the rest of society, making a living based on hunting, subsistence farming, and distilling whiskey. But this lifestyle did not survive. Industrialists in the nineteenth century waged a successful campaign to secure vast amounts of mineral rights in the mountains, dominate governmental entities in the region, and exploit the mountaineer population for wage labor. This evolution culminated in central Appalachia becoming "coal country," where the majority of residents literally lived, worked, and breathed coal. Workers were isolated in so-called "company towns," where coal companies owned every amenity and most necessities, including housing. Workers would sometimes be paid in "scrip" instead of legal money—a currency limited in its exchange and accepted only at the company store (Bell and York 2010; Caudill 1963; Eisenberg 2015; Stoll 2017).

Modern-day coal enthusiasts often point to Appalachia's role as "coal country" with pride. However, it is far from clear that the benefits of the region's coal-based economy have outweighed the costs. Harms experienced by workers are among Appalachia's most prominent sacrifices (Hendryx and Ahern 2009). From 2006 to 2011 alone, coal mines saw four major coal-mining disasters in which fifty-two miners died. Yet "in addition to these disasters, which tend to catch the public's attention, hundreds of miners have died one by one, in accidents such as collapsed roofs that do not make headline news. Between 1996 and 2005, nearly 10,000 miners died of black lung disease" (Lofaso 2011).

The environmental and public health costs in coal country have also been dramatic. Mining practices, especially the relatively recent innovations in mountaintop removal mining in which land on summits is cleared to expose underlying coal seams, have been associated with severe environmental degradation and contamination in surrounding water bodies and land areas. Public health researchers Hendryx and Ahern (2009, 543) concluded in one study that "coal mining areas fared significantly worse on all indicators compared with nonmining areas of Appalachia and/or the nation." Mortality was highest in the areas with the heaviest coal-mining activity, even "after controlling for smoking rates, rural-urban location, percent male population, supply of primary care doctors, a regional South variable, poverty, race/ethnicity, and education" (Hendryx and Ahern 2009, 545). Chronic disease was elevated in both women and men, "suggesting that the effects were not due to occupational exposure, as most coal miners are men" (Hendryx and Ahern 2009, 547).

The same study illustrated that the supposed economic development benefits of Appalachia's coal-based economy are questionable. Coal-mining areas had "the highest levels of unemployment and lowest incomes" (Hendryx and Ahern 2009, 543). Socioeconomic disadvantage is also "a powerful cause of morbidity and premature mortality" (Hendryx and Ahern 2009, 547). Hendryx and Ahern (2009) concluded that the costs of the coal-based economy—if one factors in the loss of human life—simply do not justify the supposed benefits.

As the coal industry waned into the new millennium, leaving Appalachian communities with limited alternatives and increasing despair, the supposed fracking revolution seemed, some thought, to come at just the right time. Communities, industry advocates assured them, could evolve from one natural resource to another. The coal-based economy could become the natural gas economy. Some residents hoped that the region would once again see well-paying jobs and prospects for mineral royalties, whatever the environmental and public health costs may be (e.g., Saito 2012; Taylor 2018).

Fracking did indeed come to Ohio, Pennsylvania, and West Virginia. In turn, many made substantial profits from royalty checks. Some of those profits were funneled into public initiatives, such as through impact fees imposed on developers to offset some of the development costs experienced by local governments. Some residents did benefit from new jobs brought in by the new industry (Clough and Bell 2016).

Scholars have observed the questionable equity and sustainability of these benefits, however (Clough and Bell 2016). Further, as with coal, natural gas extraction brings often underestimated or poorly understood costs, such as dramatically increased truck traffic, impacts on water, and a destabilizing influx of transient workers (Castelli 2015), along with boom-bust consequences of resource-based economies. While communities' pursuit of shale gas development might yield some winners, others stand to benefit far less, yet may also bear the costs of development. The next section turns to aspects of the legal frameworks on fracking in Ohio, Pennsylvania, and West Virginia, and their relationship with determining the winners and losers in fracking development—with a particular focus on how these regimes disadvantage the poor and the vulnerable in certain respects.

Failing to Regulate the Landman Profession in Ohio, Pennsylvania, and West Virginia

Much of the gas-rich land in Pennsylvania, Ohio, and West Virginia has been leased for drilling as of this writing. For an intense period at the beginning of the "boom," though, landowners in this region were routinely approached by company agents seeking to acquire rights to drill on their land. Throughout the region, people entered into major agreements after discussion and negotiation with these agents, who tend to be known as "landmen." Household by household, then, these discussions determined a substantial part of the economic benefits that would flow to the region from fracking. Most commonly, landowners would receive up-front payments followed by production-based royalties from allowing access to their gas. Despite the power the landmen wielded in these negotiations—to harass, coerce, or misinform residents, for example—the profession tends not to be subject to legal obligations in negotiations, such as those governing the conduct of lawyers and real estate agents in similar situations (Eisenberg 2016).

Landmen have played a role in companies' acquisition of mineral rights and other rights to access land for many decades. In Harry Caudill's 1963 exposé about Appalachian marginalization and impoverishment, *Night Comes to the Cumberlands*, the historian describes early landmen's role in transferring the rights to rich coal deposits away from Appalachian residents and to northeastern developers. Caudill characterizes landmen as "men of great guile and charm . . . courteous, pleasant and wonderful storytellers," who would spend hours wooing isolated, illiterate highlanders, ultimately convincing them to sign deeds that heavily disfavored them. The typical landman was "astute . . . a graduate of a fine college [and] thoroughly aware of the implications of the transaction and of the immense wealth which he was in the process of acquiring." The locals, meanwhile, "relied upon the agent for an explanation of the instrument's contents—contents which were to prove deadly to the welfare of generations of the mountaineer's descendants." Just over 25 percent of these instruments were signed by people who could write their names. Caudill concluded that because of the unequal bargaining power in these transactions, the "minerals were virtually given away" (Caudill 1963, 72).

Today, landmen's roles in energy development usually involve approaching landowners, negotiating leases or other agreements for mineral rights with the landowners, then monitoring lease compliance as development goes forward. Yet, modern landmen's reputation has not much improved. In some areas, they have gained notoriety for unscrupulousness, akin to the stereotype of a slick used car salesmen. They have been called "lease-hounds . . . moving through neighborhoods with military precision" (Eisenberg 2016, 163). Their reputation appears to be substantiated with common stories from landowners who say they have been coerced or misled in the lease negotiation process. The New York State attorney general's office and other landowner advocates warned landowners to be wary of landmen's high-pressure sales tactics, "including pitting neighbor against neighbor, or using arguments or tactics that may not be accurate" (Eisenberg 2016, 163). These tactics are additionally concerning because of a dramatic shortage of lawyers in rural areas, which serves as a barrier to landowners securing adequate counsel and representation (Eisenberg 2016).

Landmen's unmonitored role in lease negotiations naturally puts landowners at a disadvantage, suggesting that those who are land-rich, but cash-poor, are unlikely to receive the full potential benefit of their agreement with a company. Landowners could even be made worse off, for instance, if gas well development hurt their property value, such as by destroying the productivity of a critical water source or failing to yield the predicted amount of gas. A study by the *New York Times* examining more than 111,000 oil and gas leases from 2007 to 2011 from Texas, New York, Maryland, Ohio, Pennsylvania, and West Virginia concluded that "standard leases heavily favored companies over landowners" (Eisenberg 2016). A 2014 study by economists Timmins and Vissing (2014) also highlighted concerns about lease negotiations in Texas reflecting bargaining inequities and transaction costs, such as significant differences across race groups, leases affecting housing values, and a correlation between certain lease clauses and certain violations, suggesting that the negotiation process may affect health and safety.

Despite their power to expose people to hazardous industrial activity and their murky reputation, landmen have avoided a licensing requirement in most states throughout the history of their profession. The American Association of Professional Landmen (AAPL) is a private organization with which many landmen affiliate. To gain membership, landmen must agree to abide by the AAPL's code of ethics. The AAPL insists that self-regulation is adequate to govern the profession. Yet the AAPL's ethical code consists of only a handful of provisions, and their primary objective appears to be promoting career advancement and networking for members (Eisenberg 2016).

The modern legal landscape in Appalachia has done little to mitigate the potentially disastrous impacts of landmen's capacity for ethically dubious conduct. In West Virginia, "landmen are not licensed or subject to the jurisdiction of any board or required to have any credentials" (Eisenberg 2016, 160). The same is the case in Pennsylvania and Ohio. A failed effort in 2012 by legislators in Ohio sought to require landmen to register with the Ohio Department of Natural Resources before negotiating leases. House Bill 493 would have required landmen to use a checklist confirming that the landman had made certain disclosures, ensuring that the property owner understood the lease as well as the processes and timeline involved in oil and gas production. However, the bill died in committee (Eisenberg 2016).

If landmen do deceive or coerce landowners in lease negotiations, landowners may have common law remedies available to them. Successful claims of fraud in the inducement, duress, or unconscionability, for example, could cause a court to void a lease or other contract. In one federal Ohio case, a plaintiff "argued that his lease was unconscionable because he could not read or write, had only a fourth-grade education, and was, at the time of the transaction, both elderly (aged 76) and afflicted with a medical condition that prevented him from understanding the terms of the lease" (Eisenberg 2016, 175). In a milder example from Pennsylvania, a landowner alleged that the landmen claimed "that the landowner would never be paid a bonus of more than $25 per acre, when the landowner's neighbors were, in fact, paid more than $25 per acre" (Eisenberg 2016, 175).

Leaving vulnerable landowners with the courts as their only option is troubling, however. First, rural areas have a dramatic shortage of lawyers, so it could be difficult for a rural resident to find one. Further, going to court is expensive, and the negative effects of fracking development near someone's home could be time sensitive. A fairer, more common-sense approach to landmen's potent role in these negotiations would be to impose some level of oversight, akin to the oversight of other professionals who also negotiate high-stakes financial or land development transactions. More stringent oversight of negotiations would even the playing field to the benefit of potentially vulnerable landowners, ultimately helping fracking comport with its promise of bringing meaningful economic benefits at the regional level (Eisenberg 2016).

As of this writing, the legal landscape may be changing, at least in Ohio. In 2017, Ohio's Seventh Circuit Court of Appeals issued a decision in *Dundics v. Eric Petroleum Corp* (2017-Ohio-640, para 23), requiring landmen to be licensed as real estate agents in order to secure oil and gas leases. The court stated that Ohio's statutory definition of "real estate" in Ohio Revised Code 4735, "includ[ing] leaseholds as well as any and every interest or estate in land," did encompass landmen's activities, and therefore required professionals to have a broker's license in order to secure gas rights in "any instrument affecting oil and gas," which "necessarily affects the surface rights as well." Although other states specifically exclude oil and gas leases from real estate law, Ohio's statute does not. The Ohio Supreme Court affirmed this decision on September 25, 2018.

An effort in Pennsylvania, meanwhile, was less successful. State senator Andy Dinniman introduced Senate Bill 835 in the state legislature's 2017–2018 session to rectify the fact that, as he articulated, "Currently, land agents in Pennsylvania operate with no oversight whatsoever" (Dinniman 2017). Senate Bill 835 would have imposed requirements that landmen register with the Pennsylvania Real Estate Commission, undergo background checks, and be subject to suspension for reasons such as fraud or misrepresentation (Dinniman 2017). However, the legislation was tabled without a vote in the summer of 2018.

Unfortunately, because so many landowners have already entered into agreements with companies, even if successful, these efforts toward better oversight of landmen may have come too late to prevent the continued Appalachian story of rural landowners receiving inadequate compensation for their natural resources.

Failing to Address Environmental Injustice in Ohio, Pennsylvania, and West Virginia

In the early dialogue on fracking, proponents of the process framed fracking as a "win-win-win": it is "a safe means of creating jobs, fostering economic growth in regions hard-hit by the recession, and achieving energy security" (Fish 2011,

238). Concerns about pollution, water safety, and the long-term ramifications of disrupting ancient geological formations were dismissed as "hysterical" or "anti-science" (Eisenberg 2015, 202). Not only is fracking safe, proponents argued, but it wields the potential for regional revitalization (Eisenberg 2015).

As fracking became more widespread, environmental and health concerns did receive more public attention as studies showed that the effects of fracking are mixed (US EPA 2016; Magnani et al. 2017). Concerns about environmental justice, however, appeared to remain somewhat more marginal. An important principle of environmental justice is that communities should have a meaningful chance to engage in autonomous participation in decision-making that may stand to expose them to disproportionate environmental hazards. Environmental injustice, meanwhile, is a widespread problem: environmental hazards are disproportionately imposed on communities of color and low-income populations. Many fracking skeptics did express concerns that were not directly related to pollution, but more concerned with meaningful participation and fairness. In other words, "frequently missing from the [environment versus economy] dichotomy is the fact that the concerns of many who oppose [fracking] extend beyond the purely 'environmental,' and include concerns about issues such as 'the natural resource curse' and losing autonomy" (Eisenberg 2015, 183). One anti-fracking organization in New York stated that its opposition stemmed from the perception that "the industry has little legal accountability and uses its power to undermine democratic processes, distort science, and confuse people" while another organization stated simply, "We are not a sacrifice zone" (Eisenberg 2015, 227–28).

Fears about the potential environmental justice ramifications of fracking do appear to have played out in Ohio, Pennsylvania, and West Virginia. First, studies have established that the process of natural gas development brings with it a variety of negative externalities that are imposed upon local residents. These include "pollution, adverse health effects, erosion, seismic activity, and water contamination" (Castelli 2015, 281). Water contamination can stem from migrations of methane, fracking fluid, or sediment into aquifers. Emissions into the air at frack wells also "release methane, volatile organic compounds, nitrogen oxides, sulfur dioxide, particulate matter, n-hexane, benzene, toluene, ethylbenzene, xylene, and hydrogen sulfide, all of which are harmful to humans" (Castelli 2015, 284). Land contaminations, too, have resulted in "vegetation loss, fish kills, and other property damage" (Castelli 2015, 285). Immediate human health concerns associated with fracking have included issues ranging from rashes to respiratory conditions. However, long-term health effects are more severe. For instance, one study in Colorado analyzing approximately 125,000 births from 1996 to 2009 "found that newborns from houses closest to natural gas operations had a thirty percent increase in congenital heart conditions" (Castelli 2015, 283; McKenzie et al. 2014).

Some data suggest that these negative externalities have been disproportionately imposed on poor communities and residents. Counties that have experienced the most shale gas development activity include "some of the most impoverished counties" in certain states (Castelli 2015, 286). There is an unfortunate dearth of formal research on West Virginia and Ohio, but several studies have focused on Pennsylvania. "In Pennsylvania, only one of the seven counties with the most fracking wells has a poverty level above the state average, and only marginally so" (Castelli 2015, 287). Researchers at Clark University found a strong association between poverty levels and active fracking wells in Pennsylvania (Ogneva-Himmelberger and Huang 2015). A study by Clough and Bell (2016, 6) found "no evidence of traditional distributive environmental injustice" in Pennsylvania in that there was "not a disproportionate number of minority or low-income residents in areas near to unconventional wells." However, their study did indicate that "there is benefit sharing distributive environmental injustice" in that "the income distribution of the population living closest to shale gas wells has not been transformed" (2016, 7) since shale gas development. Clough and Bell (2016, 7) conclude that "the economic benefits of shale gas production are probably not concentrated among those living with its hazards." Some anecdotal information from lawsuits and community member complaints suggests that similar problems arise in Ohio and West Virginia (Eisenberg 2015).

One might suggest that the association between fracking and high poverty rates is to be expected. Poorer communities have a clear incentive to pursue shale gas development because they have a greater need for the economic activity. Problematically, however, it is individual private landowners who stand to benefit, while neighbors must still bear the costs, raising concerns of distributive justice (Clough and Bell 2016; Fish 2011).

The evidence of environmental justice concerns raises the question: How have these states' legal regimes failed to prevent environmental injustice, and what have they done to remedy it? As discussed in chapter 2, notably, first, the federal legal regime on fracking has been virtually nonexistent. An alternative to meaningful state-level regulation would be a standardized federal framework. Yet "while some federal laws apply to hydraulic fracturing, most contain large exceptions and only apply to narrow parts of the operation . . . leav[ing] much of the fracking process unregulated and [failing to] provide the needed legal framework to help the rural poor redress fracking's negative externalities" (Castelli 2015, 292). Specifically, fracking has largely been exempted from the Safe Drinking Water Act, the Clean Water Act, the Clean Air Act, the Resource Conservation and Recovery Act, the Comprehensive Environmental Response, Compensation, and

Liability Act of 1980, the National Environmental Policy Act, and the Emergency Planning and Community Right-to-Know Act (Castelli 2015). In the absence of federal oversight, one would hope that states would seek to provide adequate regulation of this process.

One potential factor in environmental justice concerns related to shale gas development is Appalachian states' restrictions on local government autonomy. When New York State issued its ban on fracking in 2014, the state's Department of Environmental Conservation cited as its rationale the "negative socioeconomic and community character impacts" associated with shale gas development, including "the so-called boomtown phenomenon, potential loss of agricultural land, and insufficient information to make well-informed decisions" (Eisenberg 2015, 228). New York governor Andrew Cuomo also cited health risks as a reason for the ban (Eisenberg 2015). The New York policy could be criticized as paternalistic or depriving local communities of autonomous decision-making. However, in light of these risks, it was also arguably a necessary public health measure.

Ohio, West Virginia, and Pennsylvania to a lesser extent, also limited local communities' power to make decisions over gas development. However, these states did so by mandating that shale gas development would take place according to a centralized state regulatory regime. Local regulations, such as zoning ordinances, would thus be preempted by state standards (Eisenberg 2015). The main arguments in favor of state preemption are that local regulation risks a "race to the bottom" among localities, and that development will benefit from statewide standardization (McCready 2017). Ohio and West Virginia took strong stances to limit local autonomy in their legal regimes on fracking. In Ohio, the state legislature "established a uniform statewide legislative and administrative scheme that expressly preempts local regulation of oil and gas development, and in February of 2015, the Ohio Supreme Court invalidated local bans on fracking" (Eisenberg 2015, 208). West Virginia's legal regime, too, places sole responsibility for regulating fracking with the state (Eisenberg 2015).

Pennsylvania's relationship with preemption has been slightly more complicated. As discussed in chapter 2, the state's 2012 legislation, Act 13, contained zoning provisions that limited local governments' authority to zone for natural gas development. In *Robinson Township, Washington County, Pa. v. Commonwealth of Pennsylvania*, the state Supreme Court struck down the act's zoning provisions, giving more authority back to local governments.

State preemption poses a variety of problems related to environmental justice and the ability of vulnerable populations to protect themselves. It places decision-making power with agencies that may not have the resources or political will to understand unique local circumstances and to tailor decision-making to those circumstances. It also potentially deprives local communities of the chance to protect themselves from the "industry bullying" that many report experiencing with gas companies. Further, the "race to the bottom" phenomenon can also happen among states (Castelli 2015; Eisenberg 2015). In short, by limiting local governments' decision-making power, state legal regimes risk leaving vulnerable communities all the more vulnerable in their dealings with corporate actors that have better access to information, greater resources, and the prospect of taking advantage of regional economic despair (Clough and Bell 2016; Fish 2011).

It is not necessarily obvious that local governments will inevitably minimize more harms and provide more benefits if they are the main actor to regulate fracking, as opposed to state entities. Yet, according to environmental justice principles, it is problematic for a community to lack input into such major land use developments, because it places the community at greater risk of bearing environmental harms and reaping fewer economic benefits. When adequately empowered through law, local governments have a variety of tools they can use to mitigate fracking's harms, maximize local economic benefits, and mitigate concerns of distributive allocation. Local fracking ordinances can limit pollution and other nuisances by "restricting where drilling can occur, limiting hours of operation, and requiring various mitigations and safeguards" (Foster 2018, 47). They may also seek to derive tax income from drilling activities or impose impact fees for drillers' operations (Foster 2018).

A central concern with preemption in the context of fracking in this region, in particular, is the long Appalachian tradition of state ties to industry, and the continued role oil and gas lobbyists play with state legislators. In West Virginia, an "important factor" in the state's "economic plight . . . has been the coal industry's continuing political domination of state government"; as of 2018, state legislators continue to have direct financial stakes in the oil and gas industries themselves (Kotch 2018; McGinley 2004, 44). In Ohio, a nonpartisan, nonprofit organization found in 2011 that oil and gas producers spent millions of dollars in political expenditures, including "nearly $3 million" from 2001 to 2011 "in an effort to keep drilling regulations lax" (Funk 2011). Pennsylvania's former governor Tom Corbett also received millions in oil and gas campaign contributions (Couloumbis and Navratil 2017; Lord 2015). Thus, while state preemption of local governance does bring some public benefits (Funk 2011), it also raises questions as to whether preemption is a means of transferring power from local residents to the gas industry itself.

Discussion

In some areas—perhaps all too late, given the development of the shale gas boom—the law does seem to be catching up with the need to recognize and address concerns of equity in shale gas development, particularly regarding environmental justice and injustice issues. Pennsylvania's 2013 court decision may have restored some decision-making authority to local residents, while the Ohio court's 2018 decision stands to mitigate the power of the landman profession. These developments notwithstanding, these states' failure to regulate landmen and murky experience with environmental justice appeared to stem from the failure to intervene when it would benefit landowners, such as by imposing disclosure requirements on landmen—this failure ultimately benefited gas developers—alongside actively intervening when local residents did manage to demonstrate some power and influence, such as through local zoning—again, to the benefit of gas developers. These measures have also been implemented when an unknown number of vulnerable individuals may have already experienced financial or environmental harm from fracking.

Generally, both historical and current realities offer good reason to treat with skepticism the promise that fracking will bring economic benefits to Appalachia without offsetting unevenly distributed costs. Coal country's long history as a sacrifice zone bearing disproportionate environmental hazards would ideally be countered with a new story of economic diversification, an end to industry ties with state government, and more meaningful local input into community development planning. These states' legal frameworks, however, have failed to ensure such a story. They have instead allowed residents to once again bear costs with unfairly distributed benefits, lose autonomy through state preemption, and in all likelihood, face a bust even if there have been short-term gains.

References

Bell, Shannon Elizabeth, and Richard York. 2010. "Community Economic Identity: The Coal Industry and Ideology Construction in West Virginia." *Rural Sociology* 75 (1): 111–43.

Castelli, Matthew. 2015. "Fracking and the Rural Poor: Negative Externalities, Failing Remedies, and Federal Legislation." Indiana Journal of Law and Social Equality 3 (2): 281–304.

Caudill, Harry. 1963. *Night Comes to the Cumberlands*. Ashland, KY: Little, Brown.

Clough, Emily, and Derek Bell. 2016. "Just Fracking: A Distributive Environmental Justice Analysis of Unconventional Gas Development in Pennsylvania, USA." *Environmental Research Letters* 11 (2): 025001.

Couloumbis, Angela, and Liz Navratil. 2017. "Lobbyists and Lavish Campaign Donations Aimed at Thwarting Gas Severance Tax." *Pittsburgh Post-Gazette.* http://www.post-gazette.com/news/politics-state/2017/09/30/pennsylva nia-legislature-shale-severance-tax-lobbying-campaign-contributions/ stories/201709290176.

Dinniman, Andy. 2017. "Dinniman Introduces Legislative Package on Pipe-lines." Press Release, July 24. http://www.senatordinniman.com/ dinniman-introduces-legislative-package-on-pipelines.

Eisenberg, Ann M. 2015. "Beyond Science and Hysteria: Reality and Perceptions of Environmental Justice Concerns Surrounding Marcellus and Utica Shale Gas Development." *University of Pittsburgh Law Review* 77:183–234.

Eisenberg, Ann M. 2016. "Land Shark at the Door? Why and How States Should Regu-late Landmen." *Fordham Environmental Law Review* 27:157–206.

Fish, Jared. 2011. "The Rise of Hydraulic Fracturing: A Behavioral Analysis of Land-owner Decision-Making." *Buffalo Environmental Law Journal* 19:219–68.

Foster, George K. 2018. "Community Participation in Development." *Vanderbilt Jour-nal of Transnational Law* 51:39–99.

Funk, John. 2011. "Common Cause Finds Oil, Gas Industry Spends $747 Million on Lobbying, Campaign Contributions." *Plain Dealer.* https://www.cleveland.com/ business/index.ssf/2011/11/common_cause_finds_oil_gas_ind.html.

Hendryx, Michael, and Melissa M. Ahern. 2009. "Mortality in Appalachian Coal Min-ing Regions: The Value of Statistical Life Lost." *Public Health Reports* 124 (4): 541–50.

Kotch, Alex. 2018. "A Viral Video Revealed Big Energy's Stranglehold on West Vir-ginia Politics." Vice. https://www.vice.com/en_us/article/d3wdn7/a-viral-video-revealed-big-energys-stranglehold-on-west-virginia-politics.

Lofaso, Anne Marie. 2011. "What We Owe Our Coal Miners." *Harvard Law and Policy Review* 5:87–113.

Lord, Rich. 2015. "How Pennsylvania Gas Industry Gained Corbett Influence." *Pitts-burgh Post-Gazette.* http://www.post-gazette.com/news/state/2015/07/27/ How-gas-industry-gained-Corbett-influence/stories/201507270009.

Magnani, Maria B., Michael L. Blanpied, Heather R. DeShon, and Matthew J. Horn-bach. 2017. "Discriminating between Natural versus Induced Seismicity from Long-Term Deformation History of Intraplate Faults." *Science Advances* 3 (11): e1701593.

McCready, Benjamin L. 2017. "Like It or Not, You're Fracked: Why State Preemption of Municipal Bans Are Unjustified in the Fracking Context." *Drexel Law Review* 9 (1): 61–100.

McGinley, Patrick. 2004. "From Pick and Shovel to Mountaintop Removal: Environ-mental Injustice in the Appalachian Coalfields." *Environmental Law* 34:21–106.

McKenzie, Lisa M., Ruixin Guo, Roxana Z. Witter, David A. Savitz, Lee S. Newman, and John L. Adgate. 2014. "Birth Outcomes and Maternal Residential Proximity to Natural Gas Development in Rural Colorado." *Environmental Health Perspec-tives* 122:412–17.

Ogneva-Himmelberger, Yelena, and Liyao Huang. 2015. "Spatial Distribution of Unconventional Gas Wells and Human Populations in the Marcellus Shale in the United States: Vulnerability Analysis." *Applied Geography* 60:165–74.

Saito, Mhari. 2012. "Project's Promise of Jobs Has Appalachia Seeing Stars." National Public Radio. https://www.npr.org/2012/01/12/145032971/ projects-promise-of-jobs-has-appalachia-seeing-stars.

Stoll, Steven. 2017. *Ramp Hollow: The Ordeal of Appalachia*. New York: Hill and Wang.

Taylor, James. 2018. "Closing Coal Power Plants, Replacing with Natural Gas, Makes Economic Sense." Forbes Opinion. https://www.forbes.com/sites/jamestaylor/2018/02/26/closing-coal-power-plants-replacing-with-natural-gas-makes-economic-sense/#73b127322389.

Timmins, Christopher, and Ashley Vissing. 2014. "Shale Gas Leases: Is Bargaining Efficient and What Are the Implications for Homeowners If It Is Not?" http://public.econ.duke.edu/~timmins/Timmins_Vissing_11_15.pdf.

US EPA. 2016. "Hydraulic Fracturing for Oil and Gas: Impacts from the Hydraulic Fracturing Water Cycle on Drinking Water Resources in the United States (Final Report)." Washington, DC: US Environmental Protection Agency. EPA/600/R-16/236F. https://cfpub.epa.gov/ncea/hfstudy/recordisplay.cfm?deid=332990.

FRAMING FRACKING THROUGH LOCAL LENSES

Pamela A. Mischen and Joseph T. Palka Jr.

Looking at the national news media, one might expect that municipalities with horizontal high-volume hydraulic fracturing (HVHF, or "fracking") would be the front lines for battles over legislation to ban or restrict this drilling process, with clear differences among those for and against shale gas development. This chapter uses the Marcellus Shale in Pennsylvania, where there has been a diversity of response to opportunities for shale gas development, to understand the contentious politics of HVHF in the Marcellus Shale in Pennsylvania. It uses a social constructivist approach to analyze perceptions of HVHF, specifically how local municipal officials frame this issue. Our central finding is that local government officials have a nuanced view of HVHF that encompasses both its environmental costs and economic benefits.

Much attention has been paid to public opinion polls at the national and state levels regarding support for and opposition to "fracking."[1] Overall, HVHF has been a contentious issue, and news media have highlighted the extremes of the debate. Where HVHF has occurred within densely populated areas, tensions have been notably high, as in Denton, Texas (Maqbool 2015). However, most of the gas and oil drilling in general, and HVHF in particular, has occurred in rural, sparsely populated areas. These local areas are governed by municipalities that may employ as few as a handful of people, yet they have the authority (at least in Pennsylvania) to manage HVHF through a variety of processes.

In order to set the stage for this analysis, we explain what it means to take a social constructivist approach to the analysis. We then provide context for local government officials' viewpoints by examining existing research on public opinion of "fracking" and the factors that impact those views. We explore how divergent views on HVHF might be resolved at the local level. Finally, we present evidence from municipal officials in twenty-six municipalities in Pennsylvania that illustrates how HVHF has been socially constructed and the divergent views accommodated.

Social Construction and HVHF

How do local government officials influence, represent, and reproduce the local values, attitudes, and beliefs surrounding HVHF in their communities? What happens when there are both positive and negative impacts of this process in the local community? To what extent is the local community connected to the global issues surrounding the debate over HVHF? In order to answer these questions, we take a social construction approach to this analysis. Social constructionism is an appropriate approach because local government can be viewed as the social structure that emerges from "typifications" and "recurrent patterns of interaction" (Berger and Luckmann 1966, 33). Furthermore, localities can be understood as socially constructed spaces of dependence, as well as engagement (Cox 1998). It is in these spaces that neighbors come together to find "workable solutions" to public problems (Dewey 1927), often involving a nuanced political decision-making process in which conflicting, but equally valid, viewpoints may emerge (Stone 2012). The social construction of authority (i.e., what governments at various levels should and should not be able to do) is intimately tied to the construction of place and scale as citizens debate the proper role of municipalities, counties, states, and the federal government in managing the controversial process of HVHF.

In Pennsylvania, municipalities have the authority to create drilling ordinances to exercise some control over where HVHF can occur, and some municipalities have created such ordinances. These ordinances and the informal rules that determine whether they should be developed can be viewed as forms of institutional capital (Sinclair, Mischen, and Mott 2014).

To Ostrom, institutions "can be defined as the sets of working rules that are used to determine who is eligible to make decisions in some arena, what actions are allowed or constrained, what aggregation rules will be used, what procedures must be followed, what information must or must not be provided, and what payoffs will be assigned to individuals dependent on their actions" (Ostrom 1990, 51). From a social constructionist perspective, "institutionalization occurs whenever there is a reciprocal typification of habitualized actions by types of actors"

(Ostrom 1990, 54). Ostrom and Berger and Luckmann agree that institutions and institutionalization can be formal (based in law) or informal (based in practice). What both the formal and informal institutionalization processes have in common is that they "control human conduct by setting up predefined patterns of conduct" (Berger and Luckmann 1966, 55). Consequently, as time passes, institutions are viewed as an objective rather than subjective reality because the creation of the institution predates the individual perceiving it. Studying local officials is important because they both reflect the local process of institutionalization and perpetuate it through their actions.

Public Opinion of Fracking

In order to understand the larger context in which local Pennsylvania municipal officials operate, we review the literature on public opinion of fracking as well as the factors that influence these views. A series of reports by the Center for Local, State, and Urban Policy at the University of Michigan and the Institute of Public Opinion at Muhlenberg College illustrate how national and state opinions on HVHF differ. At the national level, public opinion about HVHF is highly divided. A recent national poll clearly reflects this divisiveness (Borick and Clarke 2016). Although nearly the same percentages oppose and support "hydraulic fracturing" (39 percent and 35 percent, respectively, with 26 percent being "not sure"), 28 percent strongly oppose while 15 percent strongly support.

At the state (or provincial) level, public opinions differ and have changed over time. Lachapelle and Montpetit (2014) report data for the Province of Quebec and the states of Pennsylvania, Michigan, and New York. Of these four, residents of Pennsylvania and Michigan are more likely to support HVHF than residents of New York or Quebec. However, Michigan residents were more likely to strongly support (32 percent) than somewhat support (22 percent) while residents of Pennsylvania are more likely to somewhat support (26 percent) than strongly support (23 percent) "the extraction of natural gas from shale deposits in Michigan/Pennsylvania" in 2012 (Brown et al. 2013, 9). Residents of Pennsylvania were more likely to strongly oppose than somewhat oppose (21 percent and 19 percent, respectively), while residents of Michigan were less likely to strongly oppose than somewhat oppose (16 percent and 19 percent, respectively).

Comparing the 2012 survey to a 2014 survey that used the same question of residents in New York and Pennsylvania, the percentage of strong supporters in Pennsylvania remained the same, while the percentage of somewhat supporters increased to 31 percent (Borick, Rabe, and Lachapelle 2014). The percentages of those strongly opposing and somewhat opposing decreased over that time period. Comparing New York to Pennsylvania, New Yorkers were far more likely to strongly oppose or somewhat oppose "the extraction of natural gas from shale deposits" (27 percent and 29 percent, respectively) than Pennsylvanians (15 percent and 14 percent, respectively).

Residents of Quebec were the most likely to oppose shale gas extraction (Lachapelle and Montpetit 2014). Comparing the 2012 Pennsylvania and Michigan data to Quebec data, 36 percent of Quebec residents were strongly opposed, 31 percent somewhat opposed, 23 percent somewhat in support, and 4 percent strongly in support.

Survey data are not available at the level of the municipality and there have been few studies that have examined local variation in opinion of HVHF. In a study of Pennsylvania counties that are located in the Marcellus Shale but have different levels of HVHF activity, Brasier et al. (2011) found that those in counties with higher levels of activity had a broader awareness of both positive and negative impacts brought to the local community by Marcellus Shale development. Likewise, Arnold, Farrer, and Holahan (2018b) found that eastern Ohio landowners in active or planned drilling units overwhelmingly supported HVHF, and these opinions were informed by both economic and environmental costs and benefits. Kriesky et al. (2013) compared Washington County, PA, where there was a significant amount of HVHF in the Marcellus Shale to neighboring Allegheny County, PA, where there was no HVHF occurring. Allegheny County is home to Pennsylvania's second largest city, Pittsburgh, making it a much more urbanized county than Washington. The authors found that residents of Allegheny County were more likely to support than oppose "natural gas extraction from the Marcellus Shale," but the modal category was "neither oppose nor support" (30.1 percent). Residents of Washington County were also more likely to support than oppose HVHF. However, the modal category was "support" (26.6 percent). Those in Washington County were also more likely to state that the Marcellus Shale represented a significant economic opportunity than those in Allegheny County (47.4 percent and 34.5 percent, respectively) and more likely to state that it posed very little or no threat to the environment and public health of the region (19.2 percent and 13.4 percent, respectively).

Although not strictly comparable, these polls tell an interesting story about public opinion and proximity. At the national level, 28 percent of those surveyed strongly opposed "the extraction of natural gas and oil in the United States through the use of 'hydraulic fracturing'" in 2015 (Borick and Clarke 2016). In Pennsylvania, a state with active HVHF, the percentage that strongly opposed "the extraction of natural gas from shale deposits" fell from 21 percent in 2012 (Brown et al. 2013, 9) to 15 percent in 2014 (Borick, Rabe, and Lachapelle 2014).

In Washington County, one of the most-drilled counties in Pennsylvania, strong opposition was just 8.6 percent in 2011 (Kriesky et al. 2013). Conversely (and using the same studies), strong support was 15 percent at the national level, 23 percent in Pennsylvania, and 24.2 percent in Washington County. This seems to indicate that the closer one gets to HVHF, the more likely one is to support it.

But to draw this conclusion would be an error. Adam Briggle (2015) describes the hard-fought battle in Denton, Texas, to ban HVHF. The referendum to ban HVHF within the city limits passed with 59 percent of the votes. While not comparable for many reasons, including its size (population in 2010 of 113,383), population density (1,289 people per square mile), and the fact that it is home to two universities, it provides an important counterexample of how fracking may be perceived by local communities (US Census Bureau 2010).

Factors Influencing the Opinions of "Fracking"

Proximity

The summary above suggests that proximity to HVHF has an impact on how it is viewed. One might expect, given the much-publicized negative impacts of HVHF, that there would be evidence of the NIMBY (Not In My Back Yard) phenomenon. Evidence for NIMBYism as a result of natural gas HVHF is mixed. Studying New York municipalities and their likelihood of passing an ordinance banning HVHF, Dokshin (2016) found that distance from a HVHF well application had a curvilinear relationship with the likelihood of passing a ban, increasing with distance to about 40 miles from a potential well, decreasing from 40 to 80 miles from a potential well, and then becoming negative beyond 80 miles. It is important to note that New York State currently has a ban, and at the time of the study had a moratorium, on HVHF so proximity in this case is to a potential well site, not an actual one. The passing of local bans in New York also exhibits significant spatial autocorrelation (Hall, Schultz, and Stephenson 2018).

Alcorn, Rupp, and Graham (2017), in a study of three states with active and well-developed HVHF industries, Pennsylvania, Ohio, and Texas, investigated whether actual or perceived distance from a well increased the degree to which a respondent agreed with this statement: "Fracking should be banned until the potential dangers to human health and the environment are better understood and addressed." They found that as actual distance increased in Ohio, respondents were less likely to support a ban, while perceived "closeness" to a well in

Pennsylvania and Texas increased support for a ban. Actual distance to a well was not statistically significant in Texas or Pennsylvania.

In a nationwide study, Boudet et al. (2018) found that those who lived closer to wells were both more familiar and more supportive of "fracking." In a survey of eastern Ohio residents spread over twelve counties, Arnold, Farrer, and Holahan (2018a) found that self-reported familiarity with HVHF increased as more sources of information were consulted, though most of the sources did not help residents feel more informed. Finally, in a study of two communities with well-documented opposition to HVHF—Denton, TX, and Longmont, CO—Fisk, Park, and Mahafza (2017) found that proximity to a well blowout increased support for a ban on HVHF. However, this result did not hold for normally functioning wells. Interestingly, in Denton, as the number of wells within a half mile of the mean voting precinct population center increased, support for the ban decreased.

The Alcorn, Rupp, and Graham (2017) study raises the question of how close one has to be to HVHF for it to matter. Approximately 62 percent of those who lived within one mile of a well considered themselves "close" to a well. That percentage drops off quickly with distance with 54 percent perceiving themselves as close at a distance of 1 to 3 miles, 40 percent at 3 to 5 miles, and 24 percent at 5 to 10 miles. It is not surprising, therefore, that Davis and Fisk (2014) did not find that residing in a state in which there was HVHF occurring influenced whether respondents indicated support for "fracking," "regulation of fracking," or regulation that would require disclosure of "fracking chemicals."

The notion of perceived proximity has also been exploited by those in support of and opposition to HVHF to increase interest and authoritativeness in reporting about the Marcellus Shale HVHF (Mando 2016). Much attention has been given to these discourse coalitions, and they will be discussed in greater detail in the next section.

Party Affiliation, Ideology, and World View

Many authors have tried to explain why views about HVHF are so divergent. One of the best, and most obvious, explanations is that support of HVHF is a function of recognizing and valuing its economic advantages, and opposition to HVHF is a function of recognizing its environmental and public health disadvantages (Alcorn, Rupp, and Graham 2017; Bugden, Evensen, and Stedman 2017; Kriesky et al. 2013). Veenstra, Lyons, and Fowler-Dawson (2016) show that where one gets information about "fracking" is influenced by ideology and political party. They found that the more conservative and Republican-affiliated respondents were less likely to hold proenvironmental attitudes, less likely to get information from liberal media, and more likely to get information from conservative media, resulting in being less likely to believe that fracking poses health/environment risks.

Other studies find a relationship between ideology or political party and opinions about HVHF. Boudet et al. (2014) and Bugden, Evensen, and Stedman (2017) found support for conservatism influencing support for HVHF; Alcorn, Rupp, and Graham (2017) found that Democrats in Texas (but not Pennsylvania or Ohio) were more likely to support a ban on "fracking"; and Dokshin (2016) found that Democratic vote share increased the likelihood that a municipality would pass an "anti-fracking" ordinance in New York. However, in their study of eastern Ohio landowners with a direct stake in the HVHF process, Arnold, Farrer, and Holahan (2018b) found that party politics did not have a significant impact on how the landowners viewed HVHF.

In addition to ideology and political party, world view can also influence opinions on HVHF. One way that Lachapelle and Montpetit (2014) explain the differences in opinions on shale gas HVHF in Pennsylvania, Michigan, and Quebec is through Douglas and Wildavsky's (1982) approach to world view, which describes individuals' affinity for social interaction as well as their feelings about rules regulating behavior. From these two concepts, four world views emerge: individualism, egalitarianism, hierarchy, and fatalism. Respondents were asked to what extent they agreed with the following world views:

- Society works best when we compete as individuals (individualism).
- Government should redistribute wealth to make society more equal (egalitarianism).
- Society works best when we obey those in authority (hierarchy).
- No matter which party is in power, it's more of the same (fatalism).

Based on the percentage that strongly agreed with these statements, residents of Quebec were more likely to favor egalitarianism (42 percent), while Michigan and Pennsylvania residents most strongly agreed with individualism (37 percent and 34 percent, respectively). Residents of Quebec were least likely to agree with the individualism perspective (18 percent), residents of Michigan were least likely to agree with the egalitarian perspective (18 percent), and residents of Pennsylvania were least likely to agree with the hierarchy perspective (24 percent). Boudet et al. (2014) also show that the egalitarian world view decreased support for "fracking," but that the individualist view had no impact.

Another influencing factor on public opinions of HVHF is rural versus urban location. Dokshin (2016) found that rural areas in New York were less likely to pass an "anti-fracking" ordinance. Davis and Fisk (2014) found that urban residents were less likely to support "fracking," and that rural residents were more likely to support "fracking" and less likely to support regulation of disclosure of "fracking chemicals." Alcorn, Rupp, and Graham (2017) did not find that living in a metro area in Pennsylvania, Ohio, or Texas influenced interest in passing a "fracking" ban.

Proximity: Conflict, Moderation, or Sorting?

There are several ways that we can interpret local activism for and against HVHF. NIMBYism characterizes the response to locally unwanted land uses (LULUs). Other acronyms with negative connotations include NIABY (not in anybody's backyard), NIMTOO (not in my term of office), BANANA (build absolutely nothing anywhere near anyone), NOPE (not on planet Earth), and CAVEs (citizens against virtually everything) (Schively 2007). On the other hand, YIMBY (yes in my back yard) or PIMBY (please in my back yard, or sometimes, power in my back yard) reflect desire for development that others may reject. Dokshin (2016) found that municipalities with active landowner coalitions were less likely to pass "fracking" bans, reflecting the PIMBY phenomenon. Neville and Weinthal (2016) show how "anti-fracking" NIABY activists in Whitehorse, Yukon, used local opposition (NIMBY) to liquified natural gas (LNG) generators to insert themselves into the decision process at a local level. This had the benefit of increasing "local" opposition to the plant while broadening the NIABY coalition. An insider/outsider tension can result from outsiders inserting themselves into local politics, but also from newcomers to areas being labelled as outsiders by longer-term residents (Neville and Weinthal 2016).

Municipalities can be viewed as the physical spaces where NIMBY, PIMBY, and NIABY forces conflict. Alternatively, proximity has also been shown to moderate views (Blommaert et al. 2003). Therefore, those areas engaging in HVHF may exhibit more balanced views (i.e., seeing both the positives and negatives of HVHF) than those that are more distant. Although not studied in HVHF, the role of geographic proximity on conflict has been studied in organizational settings. Comparing distributed versus co-located work teams, Hinds and Mortensen (2005) found that co-located teams were less likely to have interpersonal conflicts. The effect of proximity on interpersonal conflict was moderated by shared identity, and shared context moderated the effect on task conflict. Furthermore, spontaneous communication had a direct moderating effect on both types of conflict suggesting that spontaneous communication helps with conflict identification and conflict handling. In network settings, Ansell and Gash (2007) argue that face-to-face dialogue allows for good-faith negotiation, which then engenders trust, and permits a virtuous cycle of collaboration to begin.

These studies can be viewed as illustrating the importance of social capital in communities. In actual geographic communities, Onyx and Bullen (2000) found that rural communities exhibited higher levels of social capital than urban communities, particularly along the lines of participation in the local community, feelings of trust and safety, and neighborhood connections. "This pattern seems to suggest that rural communities generate considerable bonding social capital, characterized by strong mutual support within the local level and high levels of participation in community life. However, such support is likely to be limited to insiders and may not be extended to minority groups within the local area or those outside the area" (Onyx and Bullen 2000, 38).

A third way to consider how differing viewpoints may be managed at the local level is to consider Tiebout sorting (Tiebout 1956). The Tiebout hypothesis is that residents can be viewed as consumer-voters who examine the local revenue-expenditure patterns and choose their community just as they choose other goods in the marketplace. In addition to choosing government services, there is evidence that people "vote with the feet" for environmental quality (Banzhaf and Walsh 2008). One implication of Tiebout's work is that municipal integration will leave people worse off because it reduces the diversity of goods in this municipal marketplace.

In sum, there are three possibilities for how divergent views on HVHF may play out at the local level. First, local communities may be the physical locations of conflict between NIMBY, PIMBY, and NIABY contingents. Second, they may be dominated by social capital and tend toward conflict mitigation and balancing of viewpoints. Third, they may avoid conflict altogether by engaging in Tiebout sorting with those supporting/opposing HVHF or favoring particular responses to it living in different communities.

Local Government Responses to HVHF in Pennsylvania

In Pennsylvania, drilling ordinances can only be put into place if a municipality has zoning. Previous work with the data used in this chapter, based on qualitative interviews with local government officials in Pennsylvania, found two general conclusions regarding zoning and resident expectations about unconventional drilling (Sinclair, Mischen, and Mott 2014; Mischen and Swim 2018). First, zoning is a contentious issue in rural Pennsylvania with some municipalities in strong support of it and others in strong opposition. These municipalities can be geographically contiguous. Mischen and Swim (2018) show the existence of zoning is the only predictor of whether or not a municipality would pass a drilling ordinance (with thirteen of the fourteen municipalities that had zoning passing such an ordinance). Second, municipal officials felt that they had the capacity to manage drilling (Sinclair, Mischen, and Mott 2014). One of the reasons that even the smallest municipalities had the capacity to manage drilling was that resident expectations regarding government intervention were very low. In these municipalities of low expectations, management meant maintaining roads and ensuring that drilling companies were paying for road maintenance. In higher expectations municipalities, management translated into passing drilling ordinances that attempted to minimize the negative impacts of drilling on those who were not benefiting economically.

These two studies provide the launching point for this analysis. We know that the zoning and nonzoning communities approach HVHF in different ways. All but one community with zoning passed a drilling ordinance, while none of the nonzoning communities were interested in managing drilling if it meant having zoning. This difference in opinions was not related to differences in how "drilling problems" were perceived, but in different expectations about local government.

Methods

We sought to understand how local government officials influence, represent, and reproduce the local values, attitudes, and beliefs surrounding HVHF by exploring how they speak about HVHF. We include in their views of HVHF their views about the process itself, the gas drilling industry, proponents and opponents to drilling, the Pennsylvania state government's role in managing drilling, their role as local officials in managing drilling, and their views of how outsiders view what is happening in their communities as related to gas drilling. Taking local officials as spokespeople for their communities, we were interested in what these narratives revealed about potential conflicts over drilling in their communities. Were these municipalities sites of conflict between NIMBY, PIMBY, and NIABY coalitions? Were they sites of compromise? Or did residents sort themselves based on their views so that drilling supporters lived in some communities and those in opposition to drilling lived in others?

In order to answer these questions, we conducted interviews with government officials from twenty-six Pennsylvania municipalities, both first- and second-class townships and boroughs, from June 2014 through January 2015. We focused on two regions in the state that had the most active drilling, but that were different in terms of their history and experience with energy extraction. The southwestern counties of Washington and Greene have had extensive oil, gas, and coal industries since the nineteenth century. The north central counties of Bradford, Tioga, and Susquehanna have not. From these counties, we selected the municipalities in two ways. First, we chose municipalities that had enacted or had considered enacting drilling-related ordinances. These municipalities were identified through a review of two local papers—the *Observer-Reporter*, which covers Washington and Greene Counties, and the *Daily Review*, which covers Bradford County. (Susquehanna and Tioga Counties did not have an equivalent daily press.) The review of the papers revealed that the use of drilling ordinances was nonexistent in Bradford County, fairly common in Washington County, and sporadic in Greene County. Second, we included in our sample municipalities with the largest number of wells, whether or not they considered drilling ordinances. This process ensured that we would include both municipalities most heavily affected by drilling as well as those that engaged in planning regarding drilling. We initially approached the top five most drilled municipalities. However, due to a low response rate in the north central region, we expanded our efforts in Susquehanna County by inviting six additional municipalities to participate.

Each of the twenty-six interviews was digitally recorded and transcribed. Interviews ranged from twenty to ninety minutes in length. We coded the data by first identifying passages that discussed views of drilling as a process and its impact, the gas drilling industry, proponents and opponents of drilling, the role of the state government of Pennsylvania in managing drilling, the role of local government in managing drilling, and mentions of "outsiders." In this study, "drilling" refers to the process of HVHF, and its economic, environmental, infrastructure, or quality-of-life impacts on local communities, while "drilling companies" refers to direct interaction and experience with the drilling and fracking industry contractors who performed the work locally. In this and the following sections, we have moved away from the term "HVHF" to the term "drilling" because it was the word most often used by our respondents. Following Miles, Huberman, and Saldaña (2014), for all of these categories except the role of the local government in managing drilling, we coded the responses in terms of valence (i.e., positive or negative view) and strength of language used (i.e., weak or strong). For example, a passage that described the "astronomical benefits" of drilling would be coded as positive strong, while a passage that described drilling as having "a negative impact on our roads" would be coded as negative weak. These coded passages were summarized to identify areas of commonality, divergence, and patterns. The passages that addressed the local role in managing drilling were also summarized in that way.

Once the data were coded in this manner, we were able to identify common and diverging narratives. Building upon Mischen and Swim (2018) and Sinclair,

Mischen, and Mott (2014), we knew that a primary difference was the existence of and opinions about zoning and the role of local governments, and this finding was central to this analysis as well. We then explored these common and divergent narratives for evidence of the three possible responses to proximity to drilling: (1) NIMBY, PIMBY, and NIABY phenomena, (2) moderation of views, and (3) Tiebout sorting.

Findings

Common Narratives

Overall, there was more agreement in perspectives on drilling than disagreement (see table 4.1). The most obvious difference was not opinions about drilling, drilling companies, opponents and proponents, the state, or outsiders, but in how the municipality should respond to drilling. Municipalities were divided into those that had drilling ordinances and felt that it was the responsibility of the municipal government to protect human well-being and the environment, and those who felt it was the responsibility of local government to protect private property rights and not manage drilling.

Local officials from every municipality in the sample agreed that drilling had been a net positive for their communities—even for one community that felt that its residents were generally opposed to drilling (this community will be discussed in greater depth in the next section). Additionally, in almost all the communities, there was recognition of both the environmental costs and economic benefits. However, the local government officials spent more time discussing the

TABLE 4.1 Common and divergent narratives

DIVERGENT NARRATIVE (MUNICIPALITY WITH ZONING)	COMMON NARRATIVES (ACROSS MUNICIPALITIES)	DIVERGENT NARRATIVES (MUNICIPALITIES WITHOUT ZONING)
	Drilling has been a positive for our community.	I favor zoning, but my
This county has been raped repeatedly.	Road damage and truck traffic are our primary concerns.	residents don't.
	We have adequate authority to manage drilling.	Maintaining roads protects the health and welfare of
	We value our water sources.	residents.

economic benefits than the environmental costs. This was done in strong and weak terms, with officials from municipalities with and without zoning speaking in positive strong terms.

Economic benefits were discussed at both the individual level and the municipal level, while environmental costs were generally discussed at the individual level. For instance, one municipal official spoke about the impact on the region as a whole: "Western Pennsylvania, if this never would happen, this area would really have been bad—bad because the housing market fell before this came about. It's been a godsend across the board."

Officials agreed that the primary areas of concern were road damage and truck traffic, viewed as affecting the community as a whole. A supervisor in a township with zoning explained, "When these trucks start running, you don't have one or two trucks going up a road. You have one hundred to two hundred trucks going up a road. You get a little country road that's used to having maybe two or three cars a day, they're not liking this. So, you have to go out and talk to these people, and then you have to talk to the industry, and you have to get them to slow down. It took a little bit of work, but we're getting a handle on it."

Although opinions about the state government's role varied from positive strong to negative strong, the majority of the municipal officials agreed that they had adequate authority and ability to manage this process because of their authority over roads and ability to create drilling ordinances (see also Sinclair, Mischen, and Mott 2014). The state was more often than not viewed as having greater capacity than municipalities to regulate drilling. However, managing where it could occur, and the use of their roads was best left up to local municipal officials. They viewed the authority that they have over the roads as a powerful mechanism to control the industry.

Additional narratives that crossed the zoning/nonzoning distinction, but that were mentioned less often, were that municipal officials valued the water sources for both drinking and for natural habitats. There was awareness that drilling had been linked to water quality and quantity issues. As one municipal official in a nonzoning township stated, "I found out going to a seminar that the water that they take, the frack water, now there's a lot of frack water that's being recycled, but certain frack water can't be recycled. So what they do is they inject that back into these old things [wells] and that water is lost forever whenever you lose that water." A municipal official from a township with zoning and a drilling ordinance was keenly aware of water quality issues and mentioned that his water was tested regularly. Another official admitted that her opinion of drilling would change dramatically if it affected the local water sources.

It is important to note a topic for which there was no common narrative—the drilling companies themselves. Drilling companies were referred to positively and negatively, and in strong and weak terms. While most of the discussions about drilling companies were about road damage and truck traffic, as indicated above, there were positive and negative stories about how drilling companies managed these issues.

Divergent Narratives

Despite these commonalities, there were several stories that did not correspond to the prevailing narratives. Two of them represented individual viewpoints. The official that had the most negative view of drilling, drilling companies, and the state of Pennsylvania spoke from a personal viewpoint when saying, "This county has been raped repeatedly." Despite this strongly negative view, she admitted that drilling had been a net positive for the community. A second official, representing a nonzoning municipality, was himself a proponent of zoning but recognized that his residents were largely opposed to it.

A third diverging narrative was also from a nonzoning municipality. This was the only interview with an official from a nonzoning municipality in which the responsibility of municipalities to protect the health and welfare of residents was invoked. However, it was done in a manner consistent with other nonzoning municipalities in that it was focused on roads. In explaining how they had the authority to shut down roads when drilling companies don't abide by the rules, one supervisor said that they were "playing the card of health and welfare of the people that live on these roads. They go in and destruct [sic] and destroy and rut them up whatever and then you come home in your little Prius and you can't get there."

Absent from the common narrative is conflict. Because of the rarity of examples relating to conflict, we treat them as divergent narratives. They are discussed in the following section, as they are examples of NIMBY, PIMBY, and NIABY conflicts, as well as one example of Tiebout sorting.

Discussion

Moderating Voices

From the discussion above, one can see that not only are there more commonalities between zoning and nonzoning municipalities, but that the commonalities reflect moderate views that recognize both the positives and negatives of drilling. Both strong positive and strong negative views of drilling and drilling companies can be found in the interviews. In eleven of the interviews, drilling was mentioned both positively and negatively, and in two of those interviews it was mentioned both strongly positively and strongly negatively (see table 4.2). The remaining fifteen municipalities spoke about drilling in positive, or positive and neutral, ways. Similarly, drilling companies were mentioned both positively and negatively in ten of the interviews, and strongly positively and strongly negatively in two of those interviews. Both zoning and nonzoning communities had positive and negative experiences with drilling companies. In nonzoning communities, comments varied from, "it was a total battle" to "we hear very positive things [from residents] about the gas companies." In zoning communities, comments about drilling companies spanned the gamut from "they have all kinds of excuses" to "they have been very responsive."

The results show that the difference in whether municipalities in the sample had zoning ordinances or not was not a determinant of their views about drilling and drilling companies. Almost all the comments by both types of municipalities related to drilling companies revolved around road issues. What these data show is that negative and positive views about drilling were not held by different subsets of the municipalities, but by a substantial portion of the sample. Opinions

TABLE 4.2 Municipal officials' opinions about drilling and drilling companies

	POSITIVE AND NEGATIVE	POSITIVE ONLY OR POSITIVE AND NEUTRAL
Drilling	11 Strongly positive and strongly negative 2	15
Drilling companies	10 Strongly positive and strongly negative 2	16

about drilling companies were more divided with some municipalities viewing drilling companies as predominantly negative, and others as predominantly positive.

In both groups, municipal officials viewed themselves as representatives of their residents. When municipal officials spoke of opponents and proponents of drilling within their municipalities, they did so neutrally, using matter-of-fact statements such as, "The only ones that you're hearing from are the ones that are involved with that particular drilling unit but, for whatever reason, don't own the mineral rights or didn't sign the lease agreement, so all of a sudden, there may be some minor opposition on their behalf." There was also general recognition that those who supported drilling were those who benefited from it most economically, and those who opposed it were generally those who did not benefit economically but had to put up with the traffic and road damage.

Finally, officials viewed the state of Pennsylvania as largely positive in terms of its role in regulating drilling. Many viewed the state as a resource that had trained personnel and knowledge that municipalities did not. As one official stated, "I doubt there is a municipality in western Pennsylvania at least that has staff, enough staff or staff with the right knowledge base to enforce what the DEP enforces. I know I certainly don't . . . so I think it's in the right hands because if it were here, if some of those environmental questions were here, it would be pointless. It would be less enforceable because people wouldn't know what they're seeing or what to deal with." Another explained, "I think the air quality and that type of thing we don't have the resources for that." There were also officials who held negative views of the state, such as "the DEP is not doing adequate monitoring, not really seriously taking a look at the issues of air pollution, water and soil pollution, and all that stuff" and "I don't think Pennsylvania realized the impact of what was here. I don't think anybody knew what was, you know what I'm saying, and I don't think that the state was prepared for it," but these views were in the minority. Unlike the views regarding drilling and drilling companies, officials did not indicate that residents had any opinion of the state's role. What was reflected in the interviews was the officials' own experiences and opinions of the state.

NIMBY, PIMBY, and NIABY Conflicts

Local officials indicated that there were more PIMBY-minded residents than NIMBY-minded ones. Within nonzoning municipalities, there were many officials who felt that their residents were of the "drill, baby, drill" mentality and that they wanted more drilling, not less, to "keep the mailbox money coming."

In these nonzoning municipalities, opinions were fairly homogeneous, with only one story of conflict (discussed as a case of Tiebout sorting below). Within municipalities with zoning, the prodrilling sentiment was not as strong. However, in one municipality with zoning, voters ousted supervisors who were trying to restrict drilling in favor of those who ran on "responsible drilling" platforms—an example of PIMBY. One of the supervisors involved told the story:

> There were a lot of games being played. A lot of the residents seen that game and they weren't happy with the direction that the township was going. By no means was it gas friendly. So, in the election in May, myself and [another candidate] ran as pro-gas candidates for responsible drilling. As long as it's done responsibly, I'm pro-gas. It has to be done responsible. We ran on a pro-gas responsible ticket, and we won with the largest margin of victory in the township history. The previous board, I unseated the seated chairman. I was 29 years old at the time. I'm young. I won with the largest margin of victory. The chairman only received 21 votes out of the whole township. I received 98 percent of the vote.

This was an issue that went beyond party politics as one of the new supervisors ran as a Democrat and the other as a Republican.

There was little evidence that there were active NIMBY groups. Only one municipality, the most heavily populated one in the sample, seemed to have had a majority of residents against drilling. We say "seemed" because a ballot referendum to ban drilling within the municipality was defeated, but only after municipal officials asked residents to defeat it, not because they were prodrilling but because passing the referendum would put them in conflict with the state regulations that a municipality must "provide for all uses" of land. When asked if the community would have supported the ban if it were legal, the manager responded that he believed that they would have.

There were two mentions of NIABY outsiders coming to municipalities to oppose drilling. One nonzoning township in the southwest region explained how it was "destroyed" financially when an outside environmental organization from the eastern part of the state came to the township and convinced them to enact an ordinance to prohibit drilling. They were taken to court and lost to the tune of "a couple of hundred thousand dollars in legal fees." In a township with zoning, during a "workshop for council" about 30 to 40 individuals were in the audience. A council person asked how many were from that township, and only one person raised their hand.

Tiebout Sorting

It is difficult to determine with the available data whether people are sorting themselves with respect to their views about local government or gas drilling. There was only one township in which a story was related to people leaving due to gas drilling. Another township experienced a significant population decline, which was attributed not to gas drilling, but to coal mining and the mining companies buying out homes above mines. What is clear in the data is there are differences in opinion about the role of local government as well as differences in what one municipal official termed "lifestyle" and "needs":

> That's one great thing about Pennsylvania, with the local government set up we're able to meet the needs of our people [with] different lifestyles. If you go, you know, southwest and northeast are pretty close to together, but in different parts of the state, people live differently. They're used to certain things in a different way. We're used to industry. We're used to coal. You go to southeast and they've got a little bit of steel out there, but you don't see many coal mines and their industry is a lot different. Their lifestyle is a lot different. By having the small divisions, the small groups like townships, it allows for us to serve the people a lot better and meet their needs.

This phenomenon is also reflected in the way officials spoke about zoning. Municipalities that had zoning and drilling ordinances spoke about them in strong positive terms, while officials in municipalities that did not have zoning spoke about it in strong negative terms. One official in a nonzoning municipality explained the position of the residents, stating, "this is our property and we'll do as we damn well please with it" and "zoning is Communistic or Socialist and they don't like it." Another said that "there's not a person, I don't think you could find 10 people in this township that would be willing to put zoning in place." On the opposite end of the spectrum, one official in a zoned municipality stated, "And planning and zoning is the only way you can protect. Out in the western end of the county . . . they have no zoning. They have no planning. It's an open wild west out there. You know, we try. At least we have a handle. We have something that we can control to a point." Another explained, "so if we can enforce the noise and the roads and the zoning locations, I think that's what we have the knowledge base to control that to help our residents." The data also suggest that local government officials are aware of their ultimate responsibility to represent their constituents. The one divergent narrative above from the official who valued zoning in a nonzoning municipality reflects that responsibility to represent.

More commonly, officials indicated that it was not only their personal opinions regarding zoning that they were offering, but that they were the dominant sentiments of their residents.

We cannot determine whether these differences are due to sorting or to the process of institutionalization, in which institutions such as zoning are viewed as an objective rather than subjective reality because the creation of the institution predates the individual perceiving it (Berger and Luckmann 1966). What is clear is that municipalities represent pockets of relative homogeneity in terms of their beliefs about local government.

A More Nuanced View of "Fracking"

Despite differences across municipalities with respect to zoning regulations, four common narratives prevailed: (1) drilling has had a net positive effect on our community, (2) road damage and truck traffic are our primary concerns with respect to drilling, (3) we have adequate authority to manage drilling activity in our municipality, and (4) we value our water resources. These common narratives reflect both the positives and negatives of the drilling process and relationships with drilling companies, as well as recognize the inherent risks of drilling to water resources.

Although "fracking" has been contentious on the national stage and in many communities in the United States, drilling in the Marcellus Shale in Pennsylvania has been relatively conflict free. While dissenting voices certainly exist, drilling has rarely been a political issue and true conflicts have been relatively minor. This can be viewed as a result of two forces. First, the roles of the municipal officials themselves in moderating conflict by recognizing the positives and negatives of HVHF, and the fact that the positives and negatives are not always borne equitably (Mischen and Swim 2018), resulted in actions that kept the peace. Second, people have varying tolerances for and values regarding the economic benefits and environmental costs and sort in such a way that the largest NIMBY contingent was confined to one municipality that at the time of the study had no drilling occurring, the PIMBY contingent dominated the nonzoning municipalities, and the areas where there was likely the most chance for NIMBY- and PIMBY-minded people to coexist occurred in municipalities with zoning and drilling ordinances that seemed to satisfy the lot. These results suggest that local officials are well-equipped to manage contentious issues such as HVHF and are possibly better equipped to make certain decisions regarding HVHF in their communities than state officials.

Note

1. We avoid using the term "fracking" because it is value-laden, preferring instead "HVHF" when we need to be specific about the process, or simply "drilling" for both readability and because that is how our interview respondents most often referred to the process. We use quotation marks to indicate the language used in other studies.

References

Alcorn, Jessica, John Rupp, and John D. Graham. 2017. "Attitudes toward 'Fracking': Perceived and Actual Geographic Proximity." *Review of Policy Research* 34:504–36.

Ansell, Chris, and Alison Gash. 2007. "Collaborative Governance in Theory and Practice." *Journal of Public Administration Research and Theory* 18:543–71.

Arnold, Gwen, Benjamin Farrer, and Robert Holahan. 2018a. "How Do Landowners Learn about High-Volume Hydraulic Fracturing? A Survey of Eastern Ohio Landowners in Active or Proposed Drilling Units." *Energy Policy* 114:455–64.

Arnold, Gwen, Benjamin Farrer, and Robert Holahan. 2018b. "Measuring Environmental and Economic Opinions about Hydraulic Fracturing: A Survey of Landowners in Active or Planned Drilling Units." *Review of Policy Research* 35 (2): 258–79.

Banzhaf, H. Spencer, and Randall P. Walsh. 2008. "Do People Vote with Their Feet? An Empirical Test of Tiebout's Mechanism." *American Economic Review* 98 (3): 843–63.

Berger, Peter L., and Thomas Luckmann. 1966. *The Social Construction of Reality: A Treatise in the Sociology of Knowledge*. New York: Anchor Books.

Blommaert, Jan, Anke Dewilde, Karen Stuyck, Katleen Peleman, and Henk Meert. 2003. "Space, Experience and Authority: Exploring Attitudes towards Refugee Centers in Belgium." *Journal of Language and Politics* 2 (2): 311 31.

Borick, Christopher, and Chris Clarke. 2016. "American Views on Fracking." Issues in Energy and Environmental Policy 28:1–7. https://papers.ssrn.com/sol3/papers.cfm?abstract_id=2781503.

Borick, Christopher, Barry G. Rabe, and Erick Lachapelle. 2014. "Public Perceptions of Shale Gas Extraction and Hydraulic Fracturing in New York and Pennsylvania." Issues in Energy and Environmental Policy 14:1–16.

Boudet, Hilary, Christopher Clarke, Dylan Bugden, Edward Maibach, Connie Roser-Renouf, and Anthony Leiserowitz. 2014. "'Fracking' Controversy and Communication: Using National Survey Data to Understand Public Perceptions of Hydraulic Fracturing." *Energy Policy* 65:57–67.

Boudet, Hilary S., Chad M. Zanocco, Peter D. Howe, and Christopher E. Clarke. 2018. "The Effect of Geographic Proximity to Unconventional Oil and Gas Development on Public Support for Hydraulic Fracturing." *Risk Analysis* 38:1871–90.

Brasier, Kathryn J., Matthew R. Filteau, Diane K. McLaughlin, Jeffrey Jacquet, Richard C. Stedman, and Timothy W. Kelsey. 2011. "Residents' Perceptions of Community and Environmental Impacts from Development of Natural Gas in the Marcellus Shale: A Comparison of Pennsylvania and New York Cases." *Journal of Rural Social Sciences* 26 (1): 32–61.

Briggle, Adam. 2015. *A Field Philosopher's Guide to Fracking*. New York: Liveright.

Brown, Erika, Kristine Hartman, Christopher Borick, Barry G. Rabe, and Thomas Ivacko. 2013. "Public Opinion on Fracking: Perspectives from Michigan and Pennsylvania." Issues in Energy and Environmental Policy 3:1–25.

Bugden, Dylan, Darrick Evensen, and Richard Stedman. 2017. "A Drill by Any Other Name: Social Representations, Framing, and Legacies of Natural Resource Extraction in the Fracking Industry." *Energy Research and Social Science* 29:62–71.

Cox, Kevin R. 1998. "Spaces of Dependence, Spaces of Engagement and the Politics of Scale, or: Looking for Local Politics." *Political Geography* 17 (1): 1–23.

Davis, Charles, and Jonathan M. Fisk. 2014. "Energy Abundance or Environmental Worries? Analyzing Public Support for Fracking in the United States." *Review of Policy Research* 31 (1): 1–16.

Dewey, John. 1927. *The Public and Its Problems: An Essay in Political Inquiry.* Chicago: Gateway Books.

Dokshin, Fedor A. 2016. "Whose Backyard and What's at Issue? Spatial and Ideological Dynamics of Local Opposition to Fracking in New York State, 2010 to 2013." *American Sociological Review* 81 (5): 921–48.

Douglas, Mary, and Aaron Wildavsky. 1982. *Risk and Culture.* Berkeley: University of California Press.

Fisk, Jonathan M., Yunmi Park, and Zachary Mahafza. 2017. "'Fractivism' in the City: Assessing Defiance at the Neighborhood Level." *State and Local Government Review* 49:105–16.

Hall, Joshua C., Christopher Schultz, and E. Frank Stephenson. 2018. "The Political Economy of Local Fracking Bans." *Journal of Economics and Finance* 42:397–408.

Hinds, Pamela J., and Mark Mortensen. 2005. "Understanding Conflict in Geographically Distributed Teams: The Moderating Effects of Shared Identity, Shared Context, and Spontaneous Communication." *Organization Science* 16 (1): 290–307.

Kriesky, J., B. D. Goldstein, K. Zell, and S. Beach. 2013. "Differing Opinions about Natural Gas Drilling in Two Adjacent Counties with Different Levels of Drilling Activity." *Energy Policy* 58:228–36.

Lachapelle, Erick, and Eric Montpetit. 2014. "Public Opinion on Hydraulic Fracturing in the Province of Quebec: A Comparison with Michigan and Pennsylvania." Issues in Energy and Environmental Policy 17:1–19.

Mando, Justin. 2016. "Constructing the Vicarious Experience of Proximity in a Marcellus Shale Public Hearing." *Environmental Communication* 10 (3): 352–64.

Maqbool, Aleem. 2015. "The Texas Town That Banned Fracking (and Lost)." BBC News, June 16. http://www.bbc.com/news/world-us-canada-33140732.

Miles, Matthew B., A. Michael Huberman, and Johnny Saldaña. 2014. *Qualitative Data Analysis.* Thousand Oaks, CA: Sage.

Mischen, Pamela A., and Stephanie Swim. 2018. "Social Equity and 'Fracking': Local Awareness and Responses." *Administration and Society.* https://doi.org/10.1177/0095399718774032.

Neville, Kate J., and Erika Weinthal. 2016. "Scaling Up Site Disputes: Strategies to Redefine 'Local' in the Fight against Fracking." *Environmental Politics* 25 (4): 569–92.

Onyx, Jenny, and Paul Bullen. 2000. "Measuring Social Capital in Five Communities." *Journal of Applied Behavioral Science* 36 (1): 23–42.

Ostrom, Elinor. 1990. *Governing the Commons: The Evolution of Institutions for Collective Action.* New York: Cambridge University Press.

Schively, Carissa. 2007. "Understanding the NIMBY and LULU Phenomena: Reassessing Our Knowledge Base and Informing Future Research." *Journal of Planning Literature* 21 (3): 255–66.

Sinclair, Thomas A. P., Pamela A. Mischen, and Rachel Mott. 2014. "The Local Implications of Hydraulic Fracturing: An Exploration of Local Government Capacity." Paper presented at the Association of Public Policy and Management National Conference, Albuquerque, NM, November 8, 2014.

Stone, Deborah A. 2012. *Policy Paradox: The Art of Political Decision Making*. New York: W. W. Norton.

Tiebout, Charles M. 1956. "A Pure Theory of Local Expenditures." *Journal of Political Economy* 64 (5): 416–24.

US Census Bureau. 2010. Census 2010. https://www.census.gov/programs-surveys/decennial-census/decade.2010.html.

Veenstra, Aaron S., Benjamin A. Lyons, and Amy Fowler-Dawson. 2016. "Conservatism vs. Conservationism: Differential Influences of Social Identities on Beliefs about Fracking." *Environmental Communication* 10 (3): 322–36.

Part II
PLANNING

USING BOOMTOWN MODELS TO UNDERSTAND THE CONSEQUENCES OF FRACKING

Adelyn Hall and Carla Chifos

Studies consistently show that economies based on natural resource extraction follow a boom and bust trend. Socioeconomic benefits, such as an increase in income, population growth, job growth, and a rise in housing prices are commonly reported (Bramlet 2013; Martin 2012; Mertz 2013; Thibodeaux 2008; West, Knipe, and Christopherson 2012; Zremski 2011). Of course, with these benefits also come drawbacks: as the area attracts outsiders, it can become a magnet for crime (Warnica 2012). Furthermore, with the infusion of wealth comes a ripple effect, raising the cost of living and forcing many established residents into economic instability (Mertz 2013).

The Marcellus Shale reserve lies beneath much of Ohio, New York, West Virginia, and Pennsylvania. In 2014, estimates stated that nearly 500 trillion cubic feet of natural gas could be extracted from the reserve ("Natural Gas" 2021) and, current consumption rates holding, Pennsylvania alone could supply up to twenty years' demand in the United States for natural gas (Spade 2013). Bradford County, Pennsylvania, once the epicenter of Marcellus Shale extraction, serves as an ideal location for a descriptive case study on socioeconomic changes when natural gas extraction occurs, as natural gas has been aggressively pursued since 2005 through the process of hydraulic fracturing (fracking). Located in the northeast quadrant of Pennsylvania, Bradford County spans 1,161 square miles. With a population of 62,000 and housing density at 25 units per square mile, the area is considered rural (Bradford County 2014).

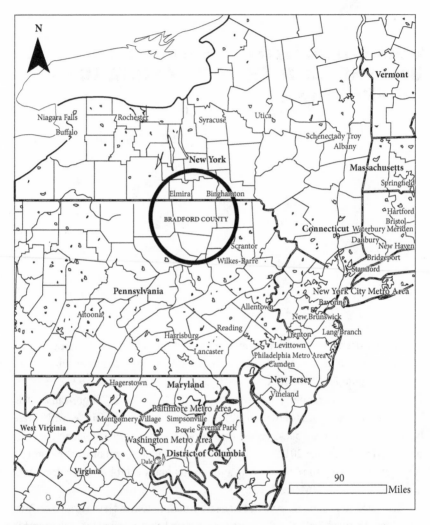

FIGURE 5.1. Location of Bradford County, Pennsylvania. Bradford County Resource Data Book 2014.

While considered fiscally depressed, from 2005 to 2014 the area underwent an economic surge due to the discovery of natural gas reserves in the Marcellus formation spanning the entire county ("Natural Gas" 2021). During this time span, Pennsylvania's Department of Environmental Protection allowed more well permits in Bradford County than any other county in the commonwealth, increasing from seven natural gas wells in 2007 to 3,021 in 2014 ("Natural Gas" 2021). During the boom of hydraulic fracking in the county, many socioeconomic changes occurred, presumably influenced by the extraction. This chapter provides a timeline of key natural gas developments in Bradford County and compares boom and bust trend models as they relate to the case study. It then examines how well each of the models explains industry impacts over time and concludes that the lack of human capital investments is key to understanding the resource curse in rural communities. This chapter captures the story of one county experiencing the boom and bust cycle of natural resource extraction. While this is not enough to generalize about all fracking locations in Pennsylvania, Bradford County provides a window into how boom and bust economies affect rural communities. Insights from this study can further inform research, as well as provide lessons that communities and their governments and planners can consider as they enter into or continue to deal with fracking in their proximity.

Boom and Bust Theories

Boom and bust theories include not only economic dynamics, such as job creation, increase in demand for services, rise of per capita income, or decrease in competitive sectors, but also environmental, political, and social aspects, such as degradation of community identity, further marginalization of disadvantaged populations, increase in traffic and air pollution, or increase in housing shortages (Albrecht 2014; Andrews and McCarthy 2014; Pender and Weber 2014; Perry 2012; Schafft, Borlu, and Glenna 2013; Spade 2013; Steele, David et al. 2012).

We use four models of boomtowns found in the literature to compare to our case study. The first is the Boomtown Model created by Gilmore in 1976, in which Gilmore outlines four stages of attitudes that boomtown communities undergo when dealing with rapid growth and industrialization. Second, the Social Disruption Boomtown Model created by Markusen in 1978 outlines six limitations faced by local governments due to natural resource extraction. Third is the Natural Gas Extraction Timeline created by Jacquet in 2010, which outlines natural gas development and breaks natural gas activity into four phases. And finally, the Resource Curse Model created by Albrecht in 2014 provides a continual boom and bust cycle that researchers state could be mitigated if investment in human capital was sustained throughout the resource extraction process. The next section of this chapter provides a closer look at these four models.

The Boomtown Model (Gilmore 1976)

By the late 1970s, the Boomtown Model emerged as a framework to describe socioeconomic changes that occur during rapid natural resource extraction, such as petroleum or coal. Due to the increase in natural gas extraction in largely rural areas, this model is reemerging as researchers hope to compare natural gas extraction patterns to that of the Boomtown Model (Heuer and Lee 2014). While the majority of researchers believe natural gas extraction will follow the same pattern of boom and bust towns, optimists state that the current technological advancements in the process will alter the boom and bust framework to allow for communities to reap the economic benefits and lower the socioeconomic risks.

The Boomtown Model comes in several different forms. The first model came in 1976, where Gilmore conceptualized a framework to explain development challenges and community attitudes in regions where natural resource extraction occurred. Gilmore described a fictional rural town that is faced with a sudden influx of population and socioeconomic activity. He describes this growth through what he calls "The Problem Triangle," in which a degraded quality of life leads to declining industrial activity, which then leads to services falling short of need, and so on.

Furthermore, Gilmore outlines four stages of attitudes that boomtown communities undergo when dealing with rapid growth and industrialization:

1. Enthusiasm
2. Uncertainty
3. Near panic
4. Adaptation

Enthusiasm describes the initial period when officials and residents concentrate on the positive economic impacts championed by the energy industry. At this time, negative impacts are dismissed or unknown. Uncertainty comes when the town starts to noticeably change and it is realized that negative impacts not only occur, but also will likely grow. Near panic comes when the industrial activity and associated impacts grow much faster than expected and services are overwhelmed. Finally, adaptation comes when the core problems are identified and mitigation strategies are developed. While residents become solidified in their beliefs, they begin to accept the reality at hand and feel a sense of progress (Gilmore 1976; Heuer and Lee 2014; Jacquet and Kay 2014).

The Social Disruption Model (Markusen 1978)

In 1978, Markusen reframed the Boomtown Model from a narrative of community attitudes to focusing on the limitations faced by local governments due to natural resource extraction. The common patterns reported are jurisdictional unevenness, conflict of residents (specifically new residents versus old residents), insufficient control of land use, rapid population growth, volatile production patterns, and monopoly of information. Further adaptation came in 1983, as Massey and Davidson, cited by Heuer and Lee in 2014, revised this model as a process of rationalization rather than a process of changing attitudes or government limitations.

The Natural Gas Extraction Timeline (Jacquet 2010)

Jacquet, following more recent natural resource extraction impacts, brings the Boomtown Model into the twenty-first century with four phases of natural gas development:

1. Predevelopment phase
2. Development phase
3. Production phase
4. Reclamation phase

The predevelopment phase is the shortest of all the phases and consists of the industry investigating the region and starting to negotiate contracts with residents for the leasing of mineral rights. During this phase, operators acquire applicable permits, conduct necessary studies and post bonds, among other job duties. The development phase is short-lived and very labor intensive. It consists of the construction of the wells and the largest influx in population due to the availability of temporary jobs. In this phase, well pad and access road construction are performed, collection of the pipeline is constructed, fracturing of the well is performed and community reclamation over some disturbances follows. The production phase is long-lived and needs only a small, but steady, labor force. This phase involves only one-tenth the workers needed in the development phase and can last for as long as natural gas is available for extraction. In this phase, trucking water and condensate from the well site occurs, as well as the monitoring of production, as production rates can decline fairly quickly. Occasionally, well workovers may be needed. Finally, the reclamation phase is where the wells are dismantled and the land is reclaimed. Figure 5.2 shows an example of projected direct employment, as well as development, production, and reclamation processes of natural gas extraction in Jonah Field, Wyoming, where Jacquet created the natural gas development model (Jacquet 2010).

The Resource Curse (Albrecht 2014)

Ultimately, sustaining rural economic development from natural gas extraction in the United States has always been a challenge; however, researchers have

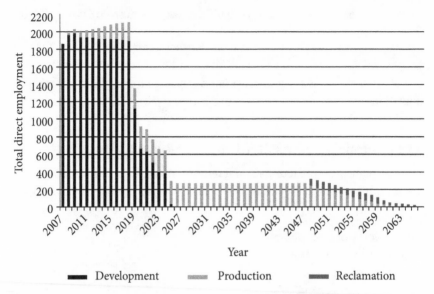

FIGURE 5.2. The expected natural gas development model in Jonah Field, Wyoming. Jacquet 2010.

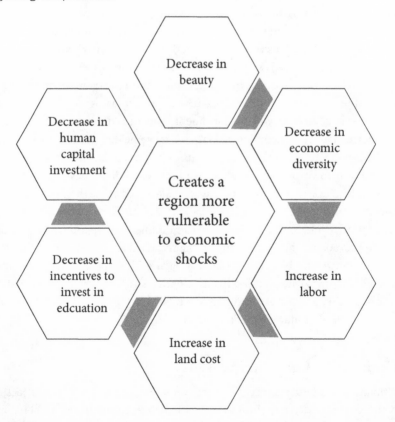

Decrease in beauty

Decrease in human capital investment

Decrease in economic diversity

Creates a region more vulnerable to economic shocks

Decrease in incentives to invest in edcuation

Increase in labor

Increase in land cost

FIGURE 5.3. The resource curse. Albrecht 2014. The figure was created by the authors from their own calculations based on the source data.

suggested best practices of resource extraction. Since the 1960s, economists have argued that investment in human and physical capital is essential for sustainable economic growth (Pender and Weber 2014). The Solow-Hartwick Rule states that a region must offset the loss in natural capital through the investment of human capital. This could be accomplished by reinvesting revenue provided by the boom into short- and longer-term costs such as infrastructure, education, public recreation, and training (Albrecht 2014; Pender and Weber 2014); however, evidence on whether revenues from natural gas development finance such investments is lacking. The resource curse provides a continual boom and bust cycle that researchers state could be mitigated if investment in human capital was sustained throughout the process of natural resource extraction (Albrecht 2014).

Natural Gas Extraction in Bradford County, Pennsylvania, 2006–2008

In 2006, the natural gas industry quietly crept into Bradford County's economy. A "silent wave" of leasing was underway as companies sought landowners to sign leases. Before 2006, some landowners had already leased their mineral rights to the oil and gas industry, selling for as little as $5.00 per acre. By 2008, Atlas Energy Resources, Rex Energy Corporation, and Range Resources had control of more than 600,000 acres of land in the Marcellus Shale play. Furthermore, by October 2008 there were 3,644 new gas leases signed in Bradford County alone, most with the oil and gas company Chesapeake Appalachia. While landowners locked in previous leases had no recourse, mineral rights began selling from $5.00 per acre to over $2,000 per acre; at its peak, mineral rights were being sold for more than $6,000 per acre. With the silent wave now turning into a frenzy, the boom of Bradford County's natural gas economy had begun (Spade 2013).

The transition from an agrarian-based economy to an industrial economy happened swiftly. Speculative economic benefits touted by state and local government, the industry, and newly leased landowners played an important role in how the residents viewed and continue to view the impacts of the gas industry, regardless of the actual spread of economic benefits. With Bradford County residents holding conservative political views, talk of the trickle-down effect through lease payments, well-paying jobs, and increased business was effective in the early days of natural gas extraction (Spade 2013). Figure 5.4 displays active permitted natural gas wells in Bradford County from 2008 through 2019. Out of

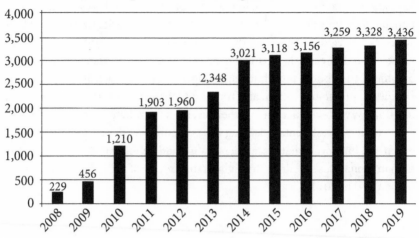

FIGURE 5.4. Active permitted natural gas wells in Bradford County, Pennsylvania. "Natural Gas" 2021. Authors compiled the data for their own calculations, from which they created this figure.

the entire Marcellus Shale region, Bradford County holds the lead in well permits issued ("Natural Gas" 2021; Marcellus Shale 2021).

2009 through 2014

Spade (2013) used qualitative research methods to gauge Bradford County's residents' range of emotions regarding the natural gas industry boom. He found that an increase in crime, increase in traffic, strains on social services, and loss of natural beauty were the main complaints expressed by most of the people he interviewed. Problems such as well contamination, increased rents, and homelessness were also expressed. Furthermore, Spade stated that the discrepancy in leasing amounts created tension and anger between neighbors and was further complicated by the fact that many leases were designed with automatic renewal clauses, not allowing for renegotiation once the lease had been signed (Spade 2013).

Positive emotions were also expressed during his investigation. The boom brought high-paying jobs, increased revenue to local businesses, and investment in infrastructure, and established new businesses in the area. With the influx of high-wage earners, rent and business increased, causing a multiplier effect for landlords and business owners, especially those in the restaurant and hotel business. Furthermore, the local government was able to increase revenue through leasing of county land, collecting fees for filing leases and collecting impact fees on natural gas through the state program, enacted in 2011.

While positive effects were observed, Spade (2013) also found undeniable unequal distribution between gas corporations, landowners, nonlandowners, residents, and outsiders that created tension and anger between neighbors. A large percentage of the jobs created by the industry went to outsiders brought in by the corporations for their expertise. Furthermore, most jobs, while high-paying, were temporary and many Bradford County locals were subsequently laid off by the gas industry as intensive construction of wells tapered off. While politicians and the industry claim economic advantages of resource extraction through the multiplier effect (which provides a means to predict job creation in an economy and is not an accounting of actual jobs created, prompting debate when discussing the theoretical versus real number of jobs supported by the oil and gas industry), most of the money generated has either been circulated outside the region or has been used to secure infrastructure that was damaged from high use by the industry, thereby negating many of the stated positive effects (Shale Tec 2012).

In US Census Bureau data, the introduction of fracking in Bradford County shows similar nuance as in Spade's research. Focusing on eight socioeconomic variables commonly affected by natural resource extraction—population, poverty, employment, industry, income, housing, rent, and crime—data collected from 2008 through 2014 show, for the most part, correlation with past research. For example, while unemployment fell and income rose during this time period, crime increased slightly, poverty increased, and home sale prices and rent increased, in some places by as much as 160 percent (Pennsylvania Uniform Crime Reporting System 2018; Shale Tec 2012; US Census Bureau 2019a). Surprisingly, local populations decreased during this time period, particularly among young adults, showing that the introduction of the natural gas industry in the area was not enough to retain this age group (Shale Tec 2012).

While the introduction of natural gas extraction produced changing economic and social dynamics in Bradford County, environmental consequences were evident as well. In April 2011, a well owned and operated by Chesapeake Energy experienced a blowout, causing tens of thousands of gallons of contaminated fluids to be spilled, causing not only a local, but a national, debate. The toxic fluid rushed into a nearby tributary, resulting in the evacuation of seven homes and the contamination of local water used for grazing. Ultimately, the Environmental Protection Agency fined Chesapeake $250,000 for the spill, the highest fine possible under current state law. Inside the county, the discussion of environmental consequences is still controversial. Some residents in the area argue that the industry has not had any adverse effects on the natural environment, while others argue that the industry is contaminating groundwater, causing habitat destruction and polluting land and streams (Spade 2013).

Emerging political discourse has seen power being stripped away from local municipalities and given to the state (Spade 2013). For example, when the oil and gas industry was introduced in Bradford County, the local conservation district was not given power over any environmental controls placed on the oil and gas industry; rather, the Pennsylvania Department of Environmental Protection governs environmental controls. While it is not necessary or wise for the local government to be the sole controller of environmental welfare, especially with industries as technical and potentially hazardous as the natural gas industry, it seems regions and states alike could benefit from more communication and enforcement overlap, lessening the feelings of dispossession in the region. Furthermore, impact fee legislation passed by former governor Corbett in 2011, which provides much-needed revenue for the region to invest in infrastructure damaged by the industry, comes with a clause that the municipalities cannot unreasonably hinder natural resource production in the area or no tax will be collected (Kulander 2013). An identical provision is placed on forestry and mining in the county (Pennsylvania Municipalities Planning Code 1968), displaying a trend toward believing that potential economic benefits continue to outweigh negative consequences of natural resource extraction.

2015 to the Present

Today, several things have changed in the oil and gas industry in Bradford County. Development has greatly slowed down in the region, first due to low gas prices caused by oversaturation of the market and second since the production phase of fracking requires minimum full-time employees. After a little more than a decade since the introduction of the industry in the area, the development phase, documented by Jacquet (2010) as the most hectic and profitable phase for the county, has nearly ended.

While natural gas extraction is no longer in its heyday in the county, in 2015 a natural gas power plant was constructed in Asylum Township, located in Bradford County. The "Panda Liberty Generating Station" is cited as the first of its kind in Pennsylvania, specifically developed to take advantage of its proximity to the Marcellus Shale gas formation. Furthermore, it is claimed to be one of the cleanest natural-gas-fueled power plants in the United States and the most efficient natural gas power plant in the world. Finally, the plant is cooled with air rather than water, saving resources in the process (Panda Power Funds 2015). This is seen as a positive step for the county, as it is beginning to invest in higher-level activities in the natural gas extraction industry. Today, the natural gas industry in Bradford County is mostly in maintenance mode, as a limited number of employees work to gauge production of the wells and conduct well workovers as needed when production rates slow. Since 2015, there have only been an additional 537 natural gas well permits approved, or a 15 percent increase in active well permits in the past five years, bringing the total active permitted gas wells to 3,436 (Marcellus Shale 2021). Further research is recommended after the release of the 2020 census in order to capture more current socioeconomic trends during this slowdown period.

Comparison of the Boom and Bust Models

As in the past with resources such as oil, timber, gold, and coal, boom and bust trends can be expected with a natural resource economy. The boom requires an influx of outside workers trained in the field and floods the area with money that flows throughout the region (Jacquet 2010). While this can be an exciting period for local governments and residents, research concludes that socioeconomic benefits have been overstated (West, Knipe, and Christopherson 2012). The bust is inevitable because gas is a natural resource and once extracted is gone and can quickly cause regret to those who decided to allow the industry in the area. While long-term jobs in the natural gas industry can be expected, it will only employ a small percentage of the total local workforce (Jacquet 2010).

Bradford County's introduction to the shale gas industry follows Gilmore's Boomtown Model. As described above, from 2006 to 2008, the industry was signing profitable leases with many landowners, and state and local governments began touting the benefits of the industry to the area. Not much was known about the impacts of the industry, but the community in general was ready to embrace oil and gas companies with open arms, going through what Gilmore refers to as a state of *enthusiasm*.

Uncertainty could be easily observed in Bradford County from 2008 to 2010, when development began in earnest in the county: many experienced long commutes, there was a changing population, impromptu employee camps were developed, and divisions in public opinion arose regarding environmental concerns. One particularly visible political division concerns the Pennsylvania and New York border, as fracking is not permitted in New York State due to environmental concerns (see chapters 2 and 4).

In 2010 and 2011, when industry activity reached its peak, *near panic* was felt in the community due to environmental hazards and cultural clashes, as many outsiders were single men and minorities from the south. Crime began to rise and many blamed the gas workers who came from outside the region. At the same time windmills were being put on the region's mountaintops, further changing the landscape of the community and greatly upsetting residents. Spade documents in his thesis "Fractured Communities" that neighbors during this time were angry at each other based on large discrepancies in their lease agreements, and social services were overwhelmed with the influx of population (2013).

Bradford County recently reached Gilmore's last state, *adaptation*, as the peak of the development came to an end and production, providing few but permanent jobs, is currently in play. The large influx of outside workers left the area and with the fall of gas prices, industry activity has slowed down more than expected. While rumored that the industry will come back in full force, it is still yet to be seen. Overall, it seems Gilmore's model accurately predicts the emotional trends communities feel when natural resource extraction occurs, but doesn't go far enough in assisting communities in mitigating the bust once industry activity comes to a stop.

The Social Disruption Boomtown Model that describes limitations faced by local governments with new natural resource extraction (Markusen 1978) is also easily related to Bradford County. As is common in rural areas, only a few municipalities have any zoning currently in place and, of those with zoning, codes are outdated and none include natural gas development.[1] Despite this major planning limitation, over time, more municipalities have worked to disseminate information to the public.

For example, in 2008 the Bradford County Planning and Mapping Services Department formed the Natural Gas Advisory Committee to study the present and future effects of the natural gas industry and disseminate information to the public. Through this committee, common topics such as drilling, well completion, transportation, and reclamation processes have been explained. Furthermore, through their community guidebook, legal rights are stated in clear language in hopes that landowners are better able to understand complicated contracts and negotiate in their favor (Bradford County 2014). Even environmental considerations, a controversial topic in the area, are examined. The guidebook concludes that best management practices need to be put into place to avoid a bust economy through investment of spin-off developments of this "vast, yet limited, resource" (Bradford County 2014).

While other regions can look to this model in order to avoid some of the stumbling blocks that Bradford County and other rural communities have had to endure, again Markusen does not provide a comprehensive model, as there are no suggestions on how communities might avoid the negative socioeconomic impacts of natural resource extraction.

Bringing the Boomtown Model into the twenty-first century, Jacquet's (2010) model closely follows Gilmore's model, but focuses on industry activity rather than community emotions. *Predevelopment* occurred in Bradford County around 2006–2008, while leases were being secured and gas companies were staking their claim. Enthusiasm was spreading and there was a collective sense that progress was coming to the sleepy county. Once the groundwork was laid, *development* started in earnest and uncertainty and panic set in. Once the infrastructure necessary to drill was mostly built and high numbers of temporary jobs led to fewer, but permanent, maintenance jobs, the community began to adapt. As the stage of *production* will ebb and flow and potentially last decades, it will be years before *reclamation* comes into play for Bradford County.

Similar to Jacquet's Expected Natural Gas Development Model in Jonah Field, Wyoming (seen in figure 5.2), figure 5.5 displays natural gas development in Bradford County, PA, from 2005 through 2019, showing active permitted natural gas wells, total direct employment of Bradford County residents, and gas production rates per year. As shown, Bradford County had a total of 3,436 active permitted natural gas wells in 2019 and hit peak production at 818,086,526,000 cubic feet in 2014 ("Natural Gas" 2021; Marcellus Shale 2021; DrillingEdge 2019). While predevelopment, development, and production has certainly been the trend in the county, the phases appear to be shorter-lived than Jacquet's model predicts. Furthermore, the US Census Bureau shows that direct employment in the mining, quarrying, and oil and gas industry peaked in 2014, with only 1,000 residents, or 4 percent of the workforce, being employed by the industry (United States Census Bureau 2019b).

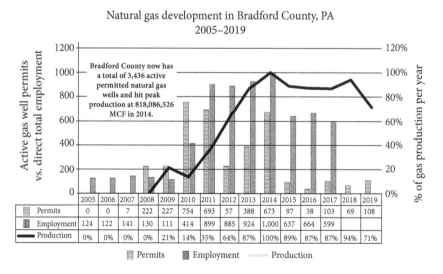

FIGURE 5.5. Natural gas development in Bradford County, Pennsylvania. Compiled by authors from multiple sources: "Natural Gas" 2021; Marcellus Shale 2021; DrillingEdge 2019; United States Census Bureau 2019b.

While Jacquet provides an accurate description of the natural resource extraction process, this model still does not answer the question of what communities can do to mitigate the negative boom and bust trends.

The Resource Curse Model by Albrecht provides some answers. As seen in Bradford County, the intense development period of the gas industry comes in waves and is unsustainable. The Bradford County Natural Gas Advisory Committee stated that other regions have endured natural gas development and felt the boom and bust cycle that seemingly follows, and the advisory committee was committed to investment in human capital in order to create a steady economy (Community Guidebook 2014).

Investments have included a commitment to outreach and training, particularly provided by local institutions, such as the Shale Training and Education Center located at the Pennsylvania College of Technology, the Penn State Marcellus Shale Research Center of Penn State University, and Cornell University's Natural Gas Resource Center, all of which have published invaluable work on natural resource extraction and continue to funnel important information to the Bradford County Natural Gas Advisory Committee. Other investments include the Panda Liberty Generating Station that was constructed in Asylum Township in 2015, as well as compressed natural gas dispensing stations that have been built in Towanda and Athens. An innovative investment from a community planning lens includes the introduction of natural gas fueled buses for the Endless Mountains Transit Authority, which provides public transportation for the entire county (Community Guidebook 2014).

Finally, Act 13 (Pennsylvania Consolidated Statutes title 58 §§ 2301–3504; see chapter 2), passed in 2012 by former governor Tom Corbett, is a law that requires municipalities that have adopted a zoning ordinance to comply with uniform standards and other requirements of the Unconventional Gas Well Impact Fee. The Well Impact Fee states, "all local ordinances regulating oil and gas operations shall allow for the reasonable development of oil and gas resources" (Community Guidebook 2014). If municipalities properly comply with these requirements, then they are issued payments from a tax on the industry to cover the costs of providing public services to the new development under thirteen specific categories as shown below in table 5.1.

As broken down in table 5.2, from 2011 to 2018 the county received a total of $49,706,260 in impact fee disbursements and Bradford County municipalities combined received a total of $77,498,575.

Finally, where municipalities are spending these funds can be an important indication of whether or not the impact fees provide the revenue and framework necessary to negate the resource curse cycle. Of the thirteen categories available, the majority of the municipalities chose to delegate their funds to the capital improvement reserve fund, which allows for municipalities to save the

TABLE 5.1 Pennsylvania impact fee acceptable usage categories

1.	Construction, reconstruction, maintenance, and repair of roadways, bridges, and public infrastructure
2.	Water, storm water, and sewer systems, including construction, reconstruction, maintenance, and repair
3.	Emergency preparedness and public safety, including law enforcement and fire services, hazardous material response, and 911 equipment
4.	Environmental programs, including trails, parks and recreation, open space, flood plain management, conservation districts, and agricultural preservation
5.	Preservation and reclamation of surface and subsurface waters and water supplies
6.	Tax reductions, including homestead exclusions
7.	Projects to increase the availability of safe and affordable housing to residents
8.	Records management, geographic information systems, and information technology
9.	The delivery of social services
10.	Judicial services
11.	Deposit into the municipality's capital reserve fund if the funds are used solely for a purpose set forth in Act 13 of 2012
12.	Career and technical centers for training workers in the oil and gas industry
13.	Local or regional planning initiatives under the act of July 31, 1968, known as the Pennsylvania Municipalities Planning Code

Source: Pennsylvania Public Utility Commission 2012.

TABLE 5.2 Impact fee disbursement, Bradford County, 2011=2018

	2011	2012	2013	2014	2015	2016	2017	2018
County disbursement	$8,428,630	$7,296,905	$7,054,000	$6,424,792	$4,923,334	$4,318,962	$5,051,256	$6,208,381
Municipal disbursement	$12,283,478	$11,147,150	$11,008,996	$10,066,514	$7,971,788	$7,029,665	$8,084,831	$9,906,153

Source: "Bradford County Impact Fee Revenue Report" 2019.

money for future expected investment in the other twelve Act 13 categories. The second most selected category was construction, reconstruction, maintenance, and repair of roadways, bridges, and public infrastructure, while emergency preparedness and records management were also commonly reported (Pennsylvania Public Utility Commission 2012). Thus far, Bradford County has allocated the majority of its impact fee dollars to infrastructure, crime and public safety, Geographic Information System (GIS) Mapping, and capital reserves. While it is not a surprise that the municipalities are choosing to focus on categories such as construction, as the wear and tear of roads from heavy vehicles and heavy traffic flows is apparent, examination of other categories not commonly selected for funds disbursements, such as environmental programs, including trails, parks and recreation, open space, flood plain management, conservation districts, and agricultural preservation, is surprising, given impacts of extraction. Whether by choice or due to lack of capacity, investment in human capital does not appear to be a top priority—not just for Bradford County—but municipalities mentioned in other chapters in this book as well. Furthermore, Act 13 is inadequate, as the current framework does not allow for investment in the broader community, such as increased education costs needed due to population growth. As the fee is only set aside to deal with degradation that the industry brings, money is not free to invest in other economic sectors such as education or healthcare, which would greatly improve the quality of life for local residents long after the natural gas industries are gone. Proof that natural resource extraction can be leveraged into making such investments are limited. Ultimately, we find Albrecht's model the most vital in educating rural communities about what is needed to avoid the negative boom and bust trends of natural resource extraction.

Key Planning Tools

As this case study and years of research demonstrate, existing communities engaged in natural gas extraction can expect to follow a boom and bust trend and be comparable to the models provided by Gilmore, Markusen, Jacquet, and Albrecht. While on this point the research is clear, academics and professionals in the planning field could offer an invaluable service to rural communities engaged in resource extraction by providing planning tools that work to build human capital in order to avoid the resource curse cycle. For example, in Blackwell, Texas, the local government has brokered a deal with a wind energy farm in the area, whereby taxes from the industry go into the local education fund (Smith 2011). Since 2005, Blackwell has been able to secure over $28 million in scholarships for its student body and every graduate receives $3,000 for each year they have spent in the district. Another example would be to utilize Act 13 dollars in long-term economic development options aimed at relieving marginalized populations, such as an increase in social services or affordable housing. Besides innovative financial investment, planners can also return to tried and true tools in order to help communities be prepared before the industry ascends, such as these:

1. Economic development planning
2. Comprehensive planning
3. Enforcement of building codes
4. Updated land use and zoning ordinances

Finally, the Bradford County Planning and Mapping Services Department being the prime example, planners can assist communities in collaborating with major universities and research centers. While this is not an exhaustive list, ultimately, these strategies would aim to put the local government and community residents in a position of power, being prepared for the consequences of the natural gas industry, rather than just planning in a reactionary state. With these tools in place, existing communities would not have to take on an unfair portion of risk from such an endeavor and the majority of residents in the community would likely experience benefits from natural resource extraction in their region. It is recommended that further research be conducted after the release of the 2020 United States Census in order to capture more current socioeconomic trends during this slowdown period, as well as examine how states reinvest oil and gas revenues for the benefit of rural communities.

Note

1. A brief discussion on the lack of zoning among municipalities in Bradford County is also in chapter 4.

References

Act 13. See Pennsylvania Consolidated Statutes title 58 §§ 2301–3504. 2012; and see chapter 2.

Albrecht, Don E., ed. 2014. *Our Energy Future: Socioeconomic Implications and Policy Options for Rural America*. London: Routledge.

Andrews, Eleanor, and James McCarthy. 2014. "Scale, Shale and the Slate: Political Ecologies and Legal Geographies of Shale Gas Development in Pennsylvania." *Journal of Environmental Studies and Sciences* 4 (1): 7–16.

Bradford County. 2014. "Bradford County 2014 Resource Data Book." Bradford County, PA.

"Bradford County Impact Fee Revenue Report." 2019. MarcellusGas.Org. Accessed December 5, 2019. https://www.marcellusgas.org/impact_fees.php?mapsize=smaller&county_id=1&muni_id=.

Bradford County Natural Gas Advisory Committee. Subcommittee for Community Planning. 2014. Developing Natural Gas Resources in Bradford County in the Heart of the Marcellus Play. "Community Guidebook." Bradford County, PA. 2014. Accessed January 27, 2014. https://bradfordcountypa.org/wp-content/uploads/2017/02/20-Community-Mapping-and-Planning-7-GuideBook-08202012.pdf.

Bramlet, Christina. 2013. "Bound for Boom or Bust?" Claims, May 2013. https://www.claimsmagdigital.com/claims/201305?pg=3#pg3.

Community Guidebook. 2014. See Bradford County Natural Gas Advisory Committee. Subcommittee for Community Planning.

DrillingEdge. 2019. "Oil and Gas Production in Bradford County, PA." DrillingEdge. Accessed December 4, 2019. http://www.drillingedge.com/pennsylvania/bradford-county.

Gilmore, J. S. 1976. "Boomtowns May Hinder Energy Resource Development: Isolated Rural Communities Cannot Handle Sudden Industrialization and Growth without Help." *Science* 191:535–40.

Heuer, Mark A., and Zui Chih Lee. 2014. "Marcellus Shale Development and the Susquehanna River: An Exploratory Analysis of Cross-Sector Attitudes on Natural Gas Hydraulic Fracturing." *Organization and Environment* 27 (1): 25–42.

Jacquet, Jeffrey. 2010. "Community and Economic Impacts of Marcellus Shale Natural Gas Development." Department of Natural Resources, Cornell University. Presentation at Cornell Cooperative Extension, May 12.

Jacquet, Jeffrey B., and David L. Kay. 2014. "The Unconventional Boomtown: Updating the Impact Model to Fit New Spatial and Temporal Scales." *Journal of Rural and Community Development* 9 (1): 1–23.

Kulander, Christopher S. 2013. "Shale Oil and Gas State Regulatory Issues and Trends." *Case Western Reserve Law Review* 63 (4): 1101–41.

Marcellus Shale. 2021. "Drilling Permits Trends January 2015–April 2021." Shale Experts. Accessed April 13, 2021. https://www.shaleexperts.com/plays/marcellus-shale/county/bradford-county-pa.

Markusen, A. 1978. "Socioeconomic Impact Models for Boomtown Planning and Policy Evaluation." Paper presented at the Western Regional Science Association, February 25, 1978.

Martin, Michel. 2012. "The Boom and Bust of Fracking." NPR. National Public Radio, December 12. Accessed January 27, 2014. http://www.npr.org/2012/12/13/167165906/the-boom-and-bust-of-fracking.

Mertz, Adam. 2013. "CUs in Boom and Bust Economies." Credit Union Magazine 79 (11): 28–33.

"Natural Gas: Bradford County, PA." Bradford County, PA. 2021. Accessed May 3, 2021. https://bradfordcountypa.org/natural-gas/.

Panda Power Funds. 2015. Panda Liberty Power Project. Accessed February 28, 2015. http://www.pandafunds.com/invest/liberty/.

Pender, John L., and Jeremy G. Weber. 2014. "Sustainable Rural Development and Wealth Creation: Five Observations Based on Emerging Energy Opportunities." *Economic Development Quarterly* 28 (1): 73–86.

Pennsylvania Consolidated Statutes title 58 §§ 2301–3504. 2012 (Act 13).

Pennsylvania Municipalities Planning Code. 1968. P.L. 805, No. 247.

Pennsylvania Public Utility Commission. 2012. "Unconventional Gas Well Fund Usage Report." Pennsylvania Public Utility Commission. Accessed March 4, 2015. http://www.puc.state.pa.us/NaturalGas/pdf/Act13_Usage_Reports/2011-Bradford_County.pdf.

Pennsylvania Uniform Crime Reporting System. Bradford County. 2018. Accessed May 3, 2021. https://www.attorneygeneral.gov/data/pennsylvania-uniform-crime-reporting-system-offenses/.

Perry, Simona L. 2012. "Environmental Reviews and Case Studies: Addressing the Societal Costs of Unconventional Oil and Gas Exploration and Production; A Framework for Evaluating Short-Term, Future and Cumulative Risks and Uncertainties of Hydrofracking." *Environmental Practice* 14 (4): 352–65.

Schafft, Kai A., Yetkin Borlu, and Leland Glenna. 2013. "The Relationship between Marcellus Shale Gas Development in Pennsylvania and Local Perceptions of Risk and Opportunity." *Rural Sociology* 78 (2): 143–66.

Shale Tec (Marcellus Shale Education and Training Center). 2011. "Pennsylvania Marcellus Shale Workforce Needs Assessment." Penn College of Technology and Penn State Extension. Williamsport: Marcellus Shale Education and Training Center.

Shale Tec (Marcellus Shale Education and Training Center). 2012. "Economic Impacts of Marcellus Shale in Bradford County: Employment and Income in 2010." Penn College of Technology and Penn State Extension. Williamsport: Marcellus Shale Education and Training Center.

Smith, Morgan. 2011. "Wind Farm Money Fuels Spending in West Texas Schools." Texas Tribune. Accessed March 29, 2015. http://www.texastribune.org/library/multimedia/wind-farm-money-spending-schools/.

Spade, Chad F. 2013. "Fractured Communities: Natural Gas, Resource Control and Social Response in Bradford County, PA." Master's thesis, University of West Virginia. https://doi.org/10.33915/etd.144.

Steele, David, Jennifer Hayes, Erika Weinthal, Robert Jackson, and Avner Vengosh. 2012. "Environmental and Social Implications of Hydraulic Fracturing and Gas Drilling in the United States: An Integrative Workshop for the Evaluation of the State of Science and Policy." Duke Environmental Law and Policy Forum. Accessed May 3, 2021. https://scholarship.law.duke.edu/cgi/viewcontent.cgi?article=1231&context=delpf.

Thibodeaux, Anna. 2008. "Boom, or Bust?" Greater Baton Rouge Business Report 26 (20) May 20: 22.

United States Census Bureau. 2019a. "Community Facts—Bradford County, PA. 2000–2019." Accessed May 3, 2021. https://data.census.gov/cedsci/all?g=0500000US42015.

United States Census Bureau. 2019b. "Industry—Bradford County, PA. 2005–2015." Accessed May 3, 2021. https://data.census.gov/cedsci/table?t=Industry&g=0500000US42015&tid=ACSST5Y2019.S2405.

Warnica, Richard. 2012. "Boom, Busts and Trouble." *Maclean's*, May 21, 2012, 32. https://archive.macleans.ca/article/2012/5/21/boom-busts-and-trouble.

West, David, Thomas Knipe, and Susan Christopherson. 2012. "Frack or Bust." Planning 78:8–13. http://www.greenchoices.cornell.edu/resources/publications/communities/Frack_or_Bust.pdf.

Willie, Matt. 2011. "Hydraulic Fracturing and 'Spotty' Regulation: Why the Federal Government Should Let States Control Unconventional Onshore Drilling." *Brigham Young University Law Review* 2011 (5): 1743–81. https://digitalcommons.law.byu.edu/lawreview/vol2011/iss5/9/.

Zremski, J. 2011. "Fracking Boom Could Go Bust in N. Y." *Buffalo News* (Buffalo, NY). Accessed January 27, 2015. https://buffalonews.com/news/fracking-boom-could-go-bust-in-n-y-while-state-debates-industry-moves-on/article_41af3f6c-bbbd-54cd-8e1b-a6d39d913c9b.html.

HYDRAULIC FRACTURING AND BOOMTOWN PLANNING IN WESTERN NORTH DAKOTA

Teresa Córdova

The Bakken oil field is being tapped and surrounding communities are experiencing the effects of large-scale resource extraction. Gas and oil extraction have occurred at such rapid rates that communities in the region are experiencing the "boomtown" phenomenon, with all its associated impacts. This chapter explores how government officials assessed and responded to this modern-day boomtown, with particular focus on the role of planning and development at the local level.

The situation in northwest North Dakota has been a subject of reporting in major national news outlets (e.g., *New York Times*, *Wall Street Journal*, CNN, National Public Radio, *National Geographic*, *Washington Post*, *Los Angeles Times*, *Harper's Magazine*), local papers throughout the state, and multiple other media outlets including YouTube (Davies 2012; Donovan 2012; Nicas 2012; Brown 2013; Madrick 2013; Mahon 2005; Manning 2013; Healy 2016; Brady 2018). Readers and listeners marveled at the accounts of housing scarcity, "man-camps," crime, the changing character of the community, and the transformation of heretofore hinterlands. Shale development in western North Dakota has also commanded the attention of academic researchers in the region (Fernando and Cooley 2016a; Fernando and Cooley 2016b; Caraher and Conway 2016), and books by journalists have added more detail (Gold 2014; Briody 2017).

Hydraulic fracturing, commonly referred to as fracking, has changed the economic, environmental—and even geopolitical—landscape of energy production, creating external forces that impinge upon local communities that are restricted in their abilities to meet the difficult challenges. Given the rapid, intense, and often disruptive impacts of the entry of external forces into a local community, the particular focus in this case study is on boomtown issues affecting Williston, North Dakota, along with planning and development responses employed by municipal officials in the region. How did local leaders, for example, attempt to (1) provide housing and infrastructure for the increased population; (2) mitigate negative impacts such as zoning and code violations, traffic, and public safety; and (3) meet social needs such as education? Did they initiate long-range planning and employ strategies for long-term stability and social cohesion?

After reviewing the various local responses, this chapter highlights what the case study reveals about the possibilities and limitation of government planning in a modern-day boomtown in light of influences and power of external and private entities. A short reflection follows on the many complex issues remaining that are not so easily solvable but need to be addressed as part of larger state and national policies related to energy development.

Methodology

The data collection process for this case study began with two trips (one in late 2011 and another in 2012) to western North Dakota, the latter of which was with a group of community and regional planning students from the University of New Mexico.[1]

These visits included site assessments; analysis of public documents; formal interviews with a range of stakeholders with a focus on public officials charged with municipal planning and land use, transportation and traffic engineering, economic development, public safety, schools, services such as libraries and social welfare; and informal conversations with residents, shop owners, and oil workers.

During the second trip, formal interviews were also conducted in the state capital with the director of the North Dakota Industrial Commission's Department of Mineral Resources, the director of the North Dakota Indian Affairs Commission, and offices of the North Dakota Office of the State Tax Commissioner and the North Dakota Petroleum Council. The traveling team combined its efforts with other students from the class who compiled data and documents.

Since 2012, additional information was compiled through follow-up phone interviews along with the use of public documents, local news accounts, and research conducted by professors from nearby universities.

The Boomtown Phenomena and the Boomtown Literature

The boomtown phenomenon reflects the accelerated, and sometimes sudden, economic and population increase in a city or town that typically overwhelms its ability to meet the various social, cultural, and infrastructure needs generated by rapid growth. It has long been an interest of boomtown researchers to understand and assess the impacts of boomtowns. With oil development in Texas in the 1920s, at least one published study described conditions similar to what was seen in northwestern North Dakota eighty years later (Chambers 1933). William Chambers described Kilgore, Texas, as an "Oil Boom Town" overflowing with people, stress, and opportunity. Public health and sanitation quickly became a problem. People were camped out or living in substandard housing, and waterways became contaminated. Providing infrastructure to keep up with the rapid influx of people and money was difficult: the small farming town that had been in decline as the price of cotton fell and the Great Depression set in now experienced a boom brought on by the discovery of oil in the area. As Chambers anticipated, after initially struggling with the boom and the inevitable bust from the decline in oil production, Kilgore would eventually benefit from the investment in infrastructure and housing stock.

Parallel to the energy boom of the 1970s, a boomtown literature burgeoned, largely in response to the fossil fuel extraction activities throughout the Rocky Mountain West and northern plains. Many of the early researchers were sociologists who paid particular attention to questions of community identity, social cohesion, and the impacts of the externally driven growth on the social and cultural fabric of communities (Cortese and Jones 1977; Cortese 1979a; Cortese 1979b; Cortese 1979c). In addition, they posed questions about community autonomy and the extent to which local communities were affected by and vulnerable to decisions made outside those communities. Cortese and Jones stated this emphatically:

> Our major point is that the focus of the student of social impacts should be directed toward structural and cultural changes that are occurring in these communities and that those communities which still have some ability to decide *whether or not* they should be impacted be made aware of the ultimate effects of that choice. (1977, 87; italics in the original)

A 1974 comparison of boom and nonboom counties in Wyoming (Kohrs 1974) described a raucous boomtown with increasing rates of divorce, crime, emergency room visits, school "delinquency" and truancy, alcoholism, depression, and attempted suicide. Throughout the 1980s, researchers (Freudenburg 1981; Cortese 1982; Summers and Branch 1984; Gulliford 1989) produced additional comparative examinations of the impacts of the energy boomtown phenomenon, demonstrating consequences that were often "disruptive," though the extent of impacts varied. Other studies described similar kinds of "social disruption" characteristics of the population boom of workers (usually single men) coming to small rural towns to earn relatively large and quick amounts of money where the pace of the influx had not yet allowed for social structures and physical infrastructure to accommodate the rapid growth (Little 1977).

Studies also examined growth management strategies and described the need for additional external resources for geographically isolated communities to tackle the impacts (Gilmore and Duff 1975). A social impact methodology emerged from some of these early studies, along with caveats in their use (Jones and Cortese 1976; Markusen 1978; Gramling and Brabant 1986). A few long-term studies suggested that after the boom—and even the inevitable bust—was over, the town's population stabilized (Smith, Krannich, and Hunter 2001).

Academics, after decades of near silence on the subject, are returning to the study of the boomtown phenomena, recognizing that shale development poses both similar and new challenges for communities facing its consequences (Jacquet and Kay 2014; Shigley 2009). Researchers are conducting various methods of impact analyses to assess economic and environmental effects of shale development (Oyakawa 2014; Nolon and Bacher 2014). In Australia, teams of boomtown researchers are producing studies on land use, transportation, and other planning issues (Weller 2009; Biermann, Olaru, and Paül 2016).

A few studies returned to the question of a local community's abilities—and rights—to have some role in deciding the nature and extent of impacts that stem from external forces related to the energy extraction industries (Manning 2014; Karl 2007). Given that extraction activities in primarily rural regions are tied to larger economic and geopolitical forces/factors, concerns about local communities' ability to determine the role and presence of shale development in their community remain salient, particularly in the face of so many counties and municipalities and even nations making the decision to ban fracking from their jurisdictions (Armstrong 2013). While a full assessment of the relationship between local communities and external forces is beyond the scope of this study, it is an example of how a local community, particularly municipal planners, might respond in the face of pressures that it did not invite. In doing so, this chapter adds to the developing literature on planning issues in modern-day boomtowns (Weller 2009; Biermann, Olaru, and Paül 2016).

The Shale Revolution and Modern-Day Boomtowns

The spike in interest by journalists and academics in boomtowns is due largely to the increases in energy production made possible from developments in the technology of gas and oil extraction from shale rocks as much as ten thousand feet beneath the surface, using combinations of sand, an array of undisclosed chemicals, and extensive amounts of water. Advances in the technology to fracture shale through both hydraulic fracturing and horizontal drilling, referred to as "fracking," created "the shale revolution" and, along with it, communities that are experiencing rapid growth associated with the rapid extraction.

The area around Williston, North Dakota, has generated particular interest because of the surrounding underground Bakken shale formation, a center of the "shale revolution" due to its size and potential for both gas and oil extraction (Piccolo 2008; US Geological Survey, n.d.). Assessments by the US Geological Society of the amount of gas and oil contained in the Bakken Formation in northwestern North Dakota, northeastern Montana, and southern Saskatchewan continue to be adjusted upward with every new release of data. Estimates of the "undiscovered, technically recoverable" gas and oil suggest that natural gas and oil extraction from this region will grow in importance, affecting national and international energy policies and the communities that lie above the underground reserves.[2]

With booms, however, there often come busts, particularly in an industry that is very sensitive to market prices balanced with costs of production. The boom, evident in 2012 in Williston, North Dakota, for example, was preceded by a boom-and-bust cycle in the 1980s that not only caught civic leaders by surprise but left them with a huge bill for infrastructure projects that were built too late to relieve the impacts and with too small a tax base when the bust came. This chapter relays accounts of how Williston officials responded to the boom of the 2010s. Did they learn from the last boom and bust cycle? Would they do it differently this time? How would they respond to the inevitable volatility of gas and oil extraction?

In the modern-day boomtown, the boom-and-bust cycles of shale extraction compound the challenges local communities face. Perhaps a study of how communities deal with these challenges can provide insights or implications transferable to other gas and oil boomtowns and to potentially any area that faces accelerated and unexpected impacts from extractive industries.

A Modern-Day Boomtown

From 1990 to 2000, Williston's population dropped from 13,131 to 12,512 (US Census Bureau 2000; Population.us 2014). By 2005, there were signs that the technology of horizontal hydraulic fracturing would soon open up gas and oil extraction in northwestern North Dakota (Pitman, Price, and LeFever 2001; US Department of the Interior 2013). By 2008, while the rest of the country was experiencing a major recession accompanied by increased unemployment, the boom in the Bakkens had taken hold. Between 2005 and 2010, Williston's population rose 2.73 percent; and in 2010, Williston's population was 14,716 with a jump to 25,586 in 2017. Williams County's population soared to over 34,000 (US Census Bureau 2017). Given the high number of people in temporary residences, these numbers are likely underestimates.

A trip to North Dakota's capital in December 2011 provided a taste of the boom along with the opportunity to delve into local accounts and engage in conversations with residents and local officials. Many lifelong residents of North Dakota were concerned about what the rapid development was going to mean for their way of life. They, too, heard or experienced stories of crime, campers parked in the Walmart parking lot, and large amounts of traffic. They also wondered what it might mean for the agricultural industry, which was an important part of North Dakota's economy, although agribusiness has taken over many of the small family farms. Promises of more jobs didn't necessarily impress them because the state already had one of the lowest unemployment rates in the country. In addition, they said, most of the new jobs were going to attract outsiders to their state, many of whom, they feared, would negatively affect their stable—and safe—way of life.

In 2012, a drive into northwestern North Dakota revealed a landscape of shale pads juxtaposed with agricultural production, trains with large black cylinders, widespread gas flares, trucks of all sizes, and, closer to Williston, more recreational vehicles than you could count. Coming into Williston, we expected the raucous and chaotic boomtown described in the media and early boomtown literature. What we found served as the basis for this chapter. Since 2012, numerous studies, primarily from North Dakota universities, have documented the many social changes facing both the newcomers to the region and longtime residents (Caraher and Conway 2016; Weber, Geigle, and Barkdull 2014). Several videos had also been produced documenting the difficulties of the change for local residents, the challenges for local officials, and the hardships for workers and their families (Prairie Public Broadcasting 2012; Veeder 2012).

While interviews of local residents revealed a positive outlook for what the changes, challenges, and future might bring, it was surely the case that the population boom had put pressure on the supply and cost of housing and had overburdened roads handling the intense truck traffic. The boom gave rise to stress on water and sewer systems, building code and zoning violations, and increased demands on schools, social services, and public safety, along with the ever-present threats of the uncertainty to the economy of the boom-and-bust cycle of energy extraction. We quickly found economic development, land use, and transportation planners hard at work addressing the many issues that were confronting their city and county. Seven years later, we can observe whether plans from 2012 were actualized and whether feelings of "we got this" are still present today.

Boom and Bust Cycles

Western North Dakota experienced a boom in the early 1980s, and then a bust. At that time, city officials welcomed the oil companies with "open arms," grateful for their presence, and demanded very little from them. When the bust came, Williams and McKenzie Counties, the cities of Williston and Watford City, and other surrounding communities bore the brunt, including financial, of the short-lived oil boom of the 1980s.

During this 1980s boom, Williston, similar to other cities in the region, floated bonds totaling $27 million to cover capital projects to meet infrastructure needs created by the boom. It took twenty years for Williston to pay off its bond debt, which caused financial strain because the bust came quickly, leaving the debt but not the revenue. In 2012, Williston's economic development team was clear that they were not making the same mistake again. In a *Wall Street Journal* video from late 2012, Gene Veeder, head of economic development for Watford City, a city just south of Williston in McKenzie County, expressed a similar sentiment:

> Last time, we greeted them with open arms. That bust happened so fast. A lot of those units didn't get built or got built . . . some communities, that were more advanced, got left with a tremendous debt—millions of dollars in debt. They paid the bill but, boy, they came with a whole different knowledge of what they were going to do this time. . . . We have said, we're not going to let you walk away and leave us with debt this time. That is one of the ways that we are protecting ourselves. So, we'll welcome you, but you are going to have to invest and be part of our communities in order to get our attention. (Veeder 2012)

As the boom took off in 2008, municipal officials were faced with the dilemma of how much capital investment they should make to meet the needs of rapid growth, without strapping future generations when the bust came. With fracking, after each well is drilled, fewer workers are needed to maintain them. In addition, any drop in oil prices juxtaposed with the costs of production would likely affect the amount and pace of oil extraction in the Bakkens. Nonetheless, the current pace of extraction and associated growth required that local and state planners anticipate growing population and service needs. Government officials must ask, "How much do you build when an oil boom may one day bust?" (Veeder 2012) and how do you build without overextending, given that a large percentage of your population could leave in a short amount of time? Predicting the population is tied to anticipating the size of the workforce, which, given the fluctuations in oil production, often defies demographic projections (Hodur and Bangsund 2013). Predicting shifts in population, however, "is a mix of art and science" (Caraher and Conway 2016, 204) and the ability of officials to meet sometimes-conflicting demands of a wide range of stakeholders must also take into account "the best understanding of current trends" (Caraher and Conway 2016, 204).

Shawn Wenko, Williston's economic development coordinator in 2012 and born and raised in Williston, received his bachelor's in hospitality administration from Black Hills State University, followed by a master's in business administration in sustainable destination management from George Washington University. This background is notable because, as he expressed to us, his hometown knowledge, combined with a focus on meeting people's service needs and his preference for sustainable development combined to shape the "art and science" of responding to Williston's growth and projected downturn. By 2014, Wenko was appointed executive director of Williston's Office of Economic Development and remains in that position today.

When we spoke to Wenko in the spring of 2016 and again in 2019, his optimism remained intact, and he was able to speak about the progress they had made in responding to the impacts on Williston of the ebbs and flows of oil production. In 2012, we noticed this optimism in nearly every local citizen that we encountered, despite the many difficult issues related to housing, the intense truck traffic, increased demands on schools and social services, public safety concerns, and the ever-present threats of economic uncertainty due to the boom-and-bust cycles of energy extraction.

Given the uncertainty of the boom-and-bust cycle of energy extraction, trying not to "make the same mistakes" is perhaps the biggest challenge to addressing the many housing, infrastructure, and service demands created by the rapid economic and population growth. Attempting to balance immediate needs with a range of possible future scenarios begins with making every effort to shift the burden of the costs to those who profit from the industrial activities—the oil companies, the developers, and the various subsidiary companies who manage the "man camps" or "crew camps." While the full description of strategies is more involved than the scope of this chapter allows, there are a number of strategies listed. (A follow-up study might more systematically assess the costs and benefits of the various strategies.) Nonetheless, we do know that there was a conscious and concerted effort on the part of Williston, Williams County, and even surrounding jurisdictions to approach the boom differently. As one bumper sticker in a surrounding county read, "Lord, give us another oil boom; we promise not to blow it again." This time, several strategies were employed to mitigate the impacts, provide infrastructure and services, ensure long-term stability, and address issues of quality of life and social cohesion.

Housing

In the time period leading up to and following our 2012 visit, nearly every media and anecdotal account of the boomtown explosion in western North Dakota referred to the imbalance between the supply of housing and the population influx. Media covered the Walmart parking lot full of camping trailers, makeshift housing units, and trailer parks on vacant lots, with additional coverage of tents in the city park, hotels filled to capacity, and, of course, "man camps," which, starting in 2010, housed temporary workers and generated media descriptions of visitors who purportedly provided more than just dancing services. Corporations such as Target Logistics, Lodging Solutions, Haliburton, and Sodexo managed the "man camps," later termed "crew camps," but by 2012, enough concerns expressed by Williston residents prompted talk of banning the facilities within city limits. Nonetheless, the rapid influx in the early stages of the boom was undoubtedly overwhelming for longtime Williston residents, newcomers, and officials.

Planning staff estimated that three hundred men lived in cars, and from school officials we learned that many students were living in twenty- to thirty-year-old trailers with no running water. Meanwhile, the costs of housing for longtime residents, particularly renters, skyrocketed. The rising housing costs were especially difficult for non-oil-industry workers, whose income did not provide them the means to pay the prices of a shifting housing market. Homeowners frequently welcomed extended family members to live with them because the housing stock was so low and the costs so high. Our interviews confirmed these housing issues, but actions were ongoing to address the housing crisis.

Local government officials were mindful of the impact on city residents of the large influx of gas and oil workers who, in some cases, brought their families. With the experience of a boom-and-bust cycle, they were also clear that companies should play a major role in providing housing for their workers and initially allowed temporary workforce housing in Williston. While temporary housing relieves some of the pressure of the rapid population influx, more permanent housing helps provide more stability for the community. The city of Williston actively sought the building of more permanent housing. Between 2011 and 2015, according to city records, building permits were issued for over 7,000 housing units or an average of 1,424 units added per year. This is in contrast to the days before the boom (24 units in 2003) and the early days of the boom (137 in 2008, 158 in 2009, and 610 in 2010).

By 2016, city policy makers wanting more permanent housing commitments chose not to renew permits for the "crew camps." The city, in response to legal challenges, allowed the companies more time to vacate, giving them until May 1, 2018, to close the temporary housing facilities and until August 1, 2018, to clean up the sites. While some companies left the area, Target Logistics continues to provide temporary housing in other parts of western North Dakota. The increase of more permanent housing stock added to local efforts to offset the impacts, not only of inadequate housing, but also on the overall municipal efforts for social cohesion in what had always been a tight-knit community.

Recognizing the importance of housing issues in response to the shale boom, Williston continues its efforts to address housing and has, in its 2017 comprehensive plan, identified several goals and strategies, many of which were already underway. As part of its goal for a diverse housing supply, the city seeks a "suitable mix" of housing types, sizes, and prices in line with appropriate zoning. The city monitors vacancy rates and the housing market, which they then share with developers, the oil industry, and other potential employers. In an effort to maintain housing quality, the city also conducts housing inspections, particularly of rental units, and has developed strategies to identify properties that need to be redeveloped or rehabilitated. Future plans include developing funding mechanisms to improve "blighted" areas. Additional goals identified in the comprehensive plan demonstrate an interest in neighborhood connectivity and beautification, including planned transitions between neighborhoods, bike paths, boulevard trees, and enforcement of weed, junk, and dilapidated housing ordinances.

Code Enforcement, Building, and Zoning

Driving into Williston in March 2012, it was easy to see that the most obvious land use violations involved the placement of campers and other makeshift housing arrangements. Mindy McEwen, a code compliance officer, confirmed that her days were busy and that much of it was spent writing up code violations and parking tickets. Some city ordinances had already been passed and others were in the works. For example, a recent ordinance prohibited big trucks and trailers from parking on city streets. In May 2012, a new ordinance declared that camper vehicles could not be parked in yards—either front or back—unless owned by the residents of the home, and even then, no one could live in the campers. Sleeping in vehicles in places like Walmart parking lots was also forbidden. Several new campgrounds were opening up to accommodate the temporary housing in campers, and city officials were coordinating the passing of the ordinance with the availability of these campgrounds so that "when we move them along, they have somewhere to go."

Meanwhile, new, more permanent housing subdivisions were being built both within Williston city limits and on either vacant or former agricultural lands surrounding the city. Doug Lalim, Williams County building official, for example, spends much of his time on reviewing plans and overseeing the permitting process where, in 2012, he saw permitting requests increase at least ten times from the previous three years. On an older map of Williams County, he showed us where new housing, commercial, and industrial developments were occurring. There are general contractors in the area from nearly every state meeting the various demands created by the influx of new residents. Lalim was particularly appreciative of the industrial, commercial, and housing development provided by Granite Peaks out of Casper, Wyoming. For example, they developed a 700-acre industrial park that was annexed into the Williston city limits, a 100-acre town center, and a multifamily and single mixed housing development. After making reference to the two previous boom and bust cycles in his lifetime, we asked him if he thought that the city/county had learned from the previous boom. His reply, "I think they are trying not to make the same mistakes."

The perspective conveyed by Lalim was evident in the Williams County Comprehensive Plan 2035 that was passed in December 2012. Explicitly acknowledging past mistakes from the previous boom and bust cycles, a primary objective of the comprehensive plan is "how best" to accommodate and plan for industrial, commercial, and residential growth. A primary vehicle for doing so is through the use of zoning to identify cost efficient growth areas, meaning those closest to existing infrastructure and developed areas. Industrial and commercial areas are identified, and incompatible uses discouraged while protecting fragile areas from development. Code enforcement, building, and zoning were useful tools employed by Williston and Williams County officials to direct the pace, type, and location of growth and development.

Truck Traffic

For every oil well in the area, District Engineer Walter Peterson, from the North Dakota Department of Transportation, estimated that there was an average of 2,500 truck trips per well hauling water, sand, and gravel, as well as the extracted oil. In 2012, the number of wells approximated 3,500; at the end of 2018, there were 12,816 wells producing oil in the Bakken region (US Geological Survey, n.d.). It is projected that over the next twenty years, there will be another 21,000 wells drilled (Dittrick 2011). Undoubtedly, the amount of truck traffic generated by each oil well places a heavy burden on roads, generating extensive wear and tear along with dangerous conditions on the highways that connect the rural towns in the region. North Dakota State University's Upper

Great Plains Transportation Institute study estimated that over $900 million would be needed over twenty years to rebuild or maintain roads heavily used by the oil industry (Dittrick 2011). Since then, it is likely that those estimates have increased.

The amount and weight of the truck traffic (at least 30 percent of the traffic) makes traffic count data to ascertain detailed traffic patterns very difficult to obtain when the rope tubes are worn so quickly. Bridges are also affected by the weight of the truck traffic, presenting additional challenges for traffic engineers. The weight of the truck traffic further affects roads during particular seasons when road weight restrictions are required. The speed of the truck traffic further exacerbates the road conditions adding the need for signals and other traffic calming measures. Within the town of Williston itself, it was startling to see huge trucks making turns on the main streets of the downtown area. Safety, wear and tear, length of local travel, driver impatience, pollution and associated health impacts are among the problems that the heavy truck traffic generated within city limits. A major project underway during our visit and since built is a truck bypass, which moves the trucks away from the downtown area. A major challenge for state and local officials, however, is how to pay for the necessary road improvements. Federal aid for roads is a source but not a solution. State funding is also necessary. In 2012, for example, the new governor dedicated $983 million to road repair.

While many of the traffic issues fell on state highways and were under state jurisdiction, at the local level, a number of traffic studies were done to assess issues and design alternatives. The temporary truck relief route and subsequent truck bypass is one of the big accomplishments of local government officials looking to address the issues of overburdened roads and intense truck traffic, and a permanent bypass is underway. In addition to major roadway and bridge improvements, by 2018, the local airport was expanded to accommodate increased flights and traffic. The city also worked with the BNSF rail company for improvements to Williston's transload facility to allow for expanded capacity to move freight.

Schools

The school population grew dramatically. According to school personnel, for example, in 1990, there were 212 kids in the high school, but in 2012 at the time of our visit, that number was over 800. From a school district official, we learned that the single biggest impact of the rapid expansion of the oil industry was the transient nature of the students. The high school, nonetheless, as a result of the new influx, had students from thirty-seven different states. Each week, the school was incorporating an average of seven to eight new students, but as an example of their transient nature when we visited the high school, there had been six new students that week, coupled with twelve who left. One female student had moved twenty-four times in the last eight years. When students move, they often lose one-half of the school year and fall further behind in their education. The housing situation affects children as well, with an estimated 2,300 homeless students in the elementary, middle, and high schools. With federal assistance available through the McKinney Vento Homelands Assistance Act, the district provides free or reduced lunch, school supplies, hygiene products, clothing, coins for laundry, and snack packs. Church groups also provide access to resources, such as a small domestic women's center and a kitchen.

Among the fifty states, North Dakota has among the highest education standards, with high graduation and proficiency rates. The transient nature and difficult living situations of the student population in Williston create a challenge for school officials, who stated that they consider it their responsibility to bring these students up to the highest educational level according to North Dakota standards and on which the state prides itself. Most of Williston's high school graduates go on to college, although with the high-paying jobs in extraction, more students are heading for the shale patches. Referring to the Norwegian roots of many longtime Williston residents, school officials described themselves as friendly, and as we observed from their interactions with one another and with students who came in and out of the office of the assistant principal, we could clearly observe the caring environment in which the educational process was occurring. It is probably no surprise, therefore, as we were told, that in that environment, there were no racial tensions and "students get along." It seems that actively integrating newcomers was a conscious strategy to address the changes and challenges in the schools.

In the schools, conscious efforts to accommodate and integrate new families became standard practice. The city financed the relocation of Williston High School, added an additional 200,000 square feet, and increased student capacity, while school bonds financed the conversion of the old high school to a middle school.

Public Safety

Residents of small towns often boast about knowing everyone in town and being able to leave their doors unlocked. We asked longtime Williston residents if they felt safe. We received a resounding yes. Nonetheless, in the period just before our visit, from 2009 to 2011, calls to the dispatch for service rose 260 percent. According to Williston police officer Amy Nickoloff, the most common complaints are related to traffic. "Traffic is huge." In the same time period, DUI incidents increased by 77 percent; criminal complaints, for example theft, by 31 percent; and assaults saw a 40 percent increase. Those numbers continued to increase so that by 2017, according to FBI data, both property and violent crime rates increased, giving Williston a much higher crime rate than 91 percent of cities of its size in North Dakota, and even above the national average of all cities with Williston being safer than only 17 percent of US cities (Neighborhood Scout 2018). Rape, though likely underreported, had numbers in 2017 that were comparable to 2012 figures. In 2015, the FBI set up an office in Williston, primarily to address the organized drug and sex trafficking, fraud, and even industrial crimes such as illegal dumping (Valentine 2015).

The Williston Police Department, the Williams County Sheriff's Department, and the State Patrol provide services to the area. Among Williston's safety initiatives, they have set up a Citizens Police Academy and have added crime prevention, communications, detective, and patrol divisions. To accommodate the population growth and new public safety challenges, a law enforcement center and fire stations were built, and the jail was expanded.

Social Services

"We're kind of getting oil boomed out," is what we heard at the Williams County Social Services office. "We have definitely seen the impacts of the oil boom in all areas," said Lois Reierson, then director of Williams County Social Services. This is where county service providers see the impacts of stressors that stem from the crowded and inadequate living conditions. The biggest issue is child protection. "Our child protection numbers have really increased," meaning that social service workers are deployed to assess and monitor potential child abuse or neglect issues and intervene when necessary. The office assists with foster care services, including the licensing of foster homes and day cares. The office also assists with access to food stamps and other financial assistance such as assistance for needy families, fuel assistance, childcare assistance, and others. As Reierson also pointed out, "not everyone is making oil field money." Julie Kyamme, director of Williston Headstart, affirms this when she points out that while there is a lot of wealth in the area with jobs with good salaries and wages, "there are still a lot of poor families." Key to the ability of the Williams County agency are the federal laws and funding that enable addressing some of the social service needs generated by the impacts of the economic activity of oil industry companies. Within the social service system, every effort was made to reach families in need by tapping into available federal and state resources. As the population in Williston stabilizes, these issues continue but are less severe.

Infrastructure

The additional population and activity in Williston placed demands on sewage and other municipal systems. The big question that emerges to meet these demands is how they will be funded. In an effort to avoid the overextension from the previous boom and bust cycles, when the city issued bonds for infrastructure projects, they did not borrow against property taxes but against gross production taxes (GPT). Other funding for quality-of-life projects, including recreation, public safety, and city expenses, came through additions to the sales tax.

Since 2012, there have been a number of infrastructure improvements, including new water treatment facilities, an expanded landfill, additional miles of water and sewer lines, and new street construction. In addition, facilities including a recreation center, a courthouse, and city administration facilities have been built or renovated.

In a 2017 document, the city of Williston laid out what had been its principles for taxation: work diligently to minimize the impact on existing residents and businesses to support industry development; allow voters to have a say in how they would like to generate the revenues required to meet the needs; provide industry support spending before general government expansion; work closely with the state to ensure understanding and support of impacts and benefits from investment at the local level with support of GPT as payment in lieu of property tax. An example of the application of these principles can be seen in the process that the Williston School District put into place to assess educational capital needs and the strategy to fund them. A task force of two dozen community members, parents, teachers, and school administrators determined a need for two elementary schools, additions to the high school, and other school improvements, and that the best way to fund these would be through general obligation bonds not to exceed $60 million and an increase in the mill levy to 20 mills. The task force recommendations were based on educational goals to reduce class size, early education, and twenty-first century learning.

Economic Development

In an effort to influence the pace and type of economic development, the economic development office established the Bakken Industrial Park with gas and oil related businesses and embarked on a promotional campaign, including bumper stickers that read "Rockin' in the Bakkens." It also created mixed use and mixed density centers to bring together commercial and residential activity. As a means to control haphazard development in the areas surrounding the city boundaries, they also created an extra territorial district. A key focus of the city's economic development office was to define economic activity with the appropriate land use, as exemplified by directing the location of industrial uses. Small business development was an important part of economic development efforts as the office sought to meet retail and service needs through small businesses. The office assisted with relocation services. Importantly, economic development officials sought a "diversified economy," which also included protection of its agricultural sector. Finally, Williston imposed impact fees on new development to help defray the cost to the city.

Long-Range Planning

The Williams County Comprehensive Plan 2035, completed in December 2012, begins with the acknowledgement that the previous boom in the late 1970s, "whose abrupt demise a few years later left many projects unfinished and placed a financial burden on the communities and individuals that only recently was overcome" (Williams County 2012, 8). Through the participation process for the plan, residents expressed their desire for a comprehensive plan that did a better job of guiding and managing the growth "to appropriate locations with the capacity to absorb the impacts of that growth in an efficient manner" (Williams County 2012, 10).

At the same time, having experienced a previous boom-bust cycle, the county is also aware that precautions need to be taken in order to avoid the negative impacts of the bust—vacancies, hollow buildings, and public debt—and to harness the boom while it lasts to maintain roads, enhance communities, retain the rural character, and leave a legacy of amenities for future generations (Williams County 2012, 10).

The extensive document lays out goals and strategies to "chart" how "best to accommodate" future growth while also addressing "the infrastructure framework necessary to support that growth: transportation, utilities, and natural resources" (Williams County 2012, 10). Interestingly, the opening section, while acknowledging private property rights, also states the importance of coordinating "individual decisions"—"at least at a basic level" (Williams County 2012, 10) as a way to mitigate the variety of potential negative impacts.

In response to the new demands of the boom, the city of Williston hired more planning staff and embarked on creating two key documents: a 2011 Annexation Study that provided direction to the city on annexing land to control the location of growth and expanding city boundaries from 4,571 acres to 12,994 acres, and the Williston-Williams County Regional Comprehensive Plan (2017), which was a revision of the 2010 comprehensive plan. The newest comprehensive plan provides an analysis of existing conditions along with detailed strategies for long-range planning.

State Government

Though a full analysis of the state government's response to the shale revolution is beyond the scope of this chapter, given that municipalities receive their enabling powers from state governments, it is an important factor in municipal responses and requires at least some contextual information. After the shale boom and bust cycle of the 1980s, the state of North Dakota was proactive on a number of fronts. Though North Dakota counties cannot tax oil revenue because of legislation from the 1950s, beginning in 1980, the state enacted a 6.5 percent extraction tax.[3] They also had the foresight to create a "legacy fund" that set aside 30 percent of extraction and production tax revenue that could not be touched until 2017 (Yardley 2016). Should there be any form of "bust," the fund could be tapped after 2017 to help state and local governments through the downturn. Efforts were also underway for state legislative action to send more of this tax revenue to affected communities like Williston. The North Dakota Tax Commission employed strategies to protect North Dakota taxpayers with excise tax set aside, a gross product tax, and the expectation that damages, such as saltwater damages to a property owner's land, would be repaid by the oil company. Conservation easements were designed to protect the safety of residential areas. Workers in the oil industry pay North Dakota taxes, and sales tax compliance officers roam the state. A resource that is available to business start-ups is the Bank of North Dakota, a state-owned bank created by the state legislature in 1919 for the purpose of protecting and enhancing agriculture, commerce, and industry in the state.

The North Dakota Department of Mineral Resources, Gas and Oil Division, in an effort to minimize the impact of drilling on the above ground landscape, regulated the size and shape of drilling pads. This division, which regulates gas and oil, also influences the creation of state laws and regulations including new rules for oil pits that ban the dumping of liquid drilling wastes into any open pit (with exceptions). Concerns for potential spills from truck accidents led to requirements of partial refinement for less volatility. For those concerned with climate change, the escape of gas via flares is of paramount importance. In November 2019, the division sponsored a public hearing "to consider natural gas capture strategies and regulatory clarity regarding gas gathering agreements." The North Dakota State Water Commission regulates water well construction and monitors drought conditions.

Whether state oversight to protect the short- and long-term interests of North Dakota residents is adequate, however, has been a matter of contention. *New York Times* articles, for example, have exposed questions about state oversight of the powerful oil industry exposing "the downside of the boom: North Dakota took on the oversight of a multibillion-dollar oil industry with a regulatory system built on trust, warnings, and second chances" (Sontag and Gebeloff 2014). In July 2019, a group of landowners in northwest North Dakota filed a suit against the state to contest a recently passed Pore Space Law (Senate Bill 2344), which gives mineral developers reign over the pore space below their property. Stating that they "have always tried to work with industry," they argue that this law constitutes an unconstitutional taking (Northwest Landowners Association 2019). Their unsuccessful attempt thus far to stop the law exemplifies what could be considered evidence of growing concerns about the state's oversight of the industry.

While this issue is not directly related to Williston's case study, it is important to note that shale development in western North Dakota affects the region's Native American communities. While the oil boom brought some benefit to the Three Affiliated Tribes, it has also meant a rise in the cost of living, particularly housing. In the area around New Town, for example, tribal member residents were "kicked out" of nontribal housing to make room for workers in the oil industry. Spillover effects from the industry came from trucks hauling waste saltwater, which, according to tribal police, were often leaking, with oil spills reportedly occurring two to three times per week. Damages to roads from the heavy truck traffic were noticeable, and the dust emissions from trucks hauling the scoria (red rock, hard clay) used in the oil pads and gravel roads seemed to be causing respiratory illnesses and also affected crop production. Women and girls were often harassed. An important policy issue was related to taxes. "How do you create a fair system on the reservation?" is a question that the North Dakota Indian Affairs Commission was attempting to answer. Similar to the Legacy Fund, a percentage of oil revenue was being set aside for a tribal endowment. After our trip, we learned about an initiative by the EPA's Office of Environmental Justice to collaborate with the Three Affiliated Tribes government on a training workshop about collaborative problem solving and how the approach may be useful in addressing issues from nearby fracking operations. This collaborative problem-solving approach suggested an alternative to opening the gates to a community without conditions (Gogal 2014). However, in 2016 to 2017, the struggles in Standing Rock to stop the Dakota Access Pipeline garnered national and international support and revealed the larger issues facing native populations in North Dakota as a result of fracking and extraction industries. Indigenous organizers who live in communities impacted by the Dakota Access and Line 3 continue their efforts to #ShutdownDAPL and #StopLine3.

And Then Came a Bust . . . or Was It?

In 2014, as predicted, oil prices dropped (Russell 2016), oil production slowed, and a "bust" was declared as national newspapers reported on a region that had pumped itself up during the boom, but now faced a bust (Yardley 2016). CNN declared an "Oil Boomtown on the Brink" (CNN Business 2015), and the *New York Times* described "an emptier feeling" brought on by the drop to $30/barrel (down from $90 or $100 just a few months earlier). "New drilling has dried up, and the flood of wealth and workers is ebbing" (Healy 2016).

However, in a follow-up interview by this author with Shawn Wenko, Williston's economic development director, in March 2016, he declared, "It's not a bust." Perhaps because they anticipated and prepared for it, he instead defined it as a "downturn" that enabled them to catch their breath. "Williston is very much open for business," he said. It is "not the gloom and doom painted in the media." Yes, "it is a slowdown, but not a bust." Sales tax revenue, for example, dropped and then quickly rebounded. In 2018, taxable purchases were $20.24 billion, which was a 12.5 percent increase from the previous year. Williston's decision to borrow against sales tax revenue rather than property values was a key ingredient to Williston's success in weathering the fluctuations of a boom-and-bust cycle.

Wenko went on to say, "We are not in period of 'rapid growth'—but 'normal growth'—we now have time to work on things." Oil production is down but pipeline activity remains.[4] Importantly, the slowdown "allows us to diversify our economy and strengthen our agricultural sector and our commercial and retail." He explained that much of the transient workforce left, but many remain, including families. He went on to describe the many social benefits that were now available to Williston residents, including a new high school, scholarship programs to attend community college, and the growth of the library, which had become a "mecca" for users of the internet. Wenko described a community that had embraced and integrated "newcomers." He stated, "so many people have come here from somewhere else. There was a bond; they were each other's community. Similar to old Williston, where people helped each other out, these people have developed their surrogate family." Wenko proudly spoke of the "community spirit" evidenced by community clean-ups and other community events and activities.

How did Williston anticipate and survive the downturn? Among the strategies that seemed to help were awareness of boom/bust cycles; good planning and economic development; proactive local and state regulations; and most importantly, shifting at least some of the cost burdens to help mitigate the cycles of boom and bust of rapid gas and oil extraction, while still reaping economic and social benefits. Since 2016, the price of crude oil has dropped and risen and dropped again. In BloombergQuint, March 2018, the headline of an article that heavily quoted Shawn Wenko declared that the "Oil Bust Was Actually Good for a Town Called 'Shale City'" (Schulz 2018). In November 2018, in a segment on NPR's *All Things Considered* (Brady 2018), North Dakota was described as "growing into its oil boom." After a December 2018 dip to $50 per barrel, in March 2019, the price was up again to $65.95. Though seemingly contradictory, fluctuations in crude oil prices are a constant. With the "wild ride" of the oil market as a given, the forward thinking local and state government officials anticipate and plan for the fluctuations. No doubt, a sustained drop in oil prices could also create long-term economic effects and really create "the bust" conditions. However, there are ample projections that, for many years to come, oil and gas extraction from shale rock is in its early stages, and much more is likely to come (Springer 2018). In Williston, for the most part, municipal leaders viewed this as good news. In 2019, Wenko refers to the shift "from a boom, to a business model," with a more permanent style of development (Shawn Wenko, interview by this author, March 25, 2019). With a vibrant economy, more families and more young people are moving to Williston.

In the 2018 launch of Williston's Economic Development podcast series, in a conversation with executive director Shawn Wenko, aviation director Anthony Dudas says, "It's been an amazing journey for my wife and I since 2011 to come to the city of Williston and have the opportunities that would not have occurred anywhere else in the country. We've made this our home. We're homeowners here. We absolutely love living in this community and seeing its leadership's commitment to enhance the quality of life, quality of place here. It's amazing to see and truly a blessing to be part of."

Shawn responds, "You know I always tell people that once you get boots on the ground, there is something about Williston that really pulls you in. Seeing the community from afar is much different than what you experience when you get here." Building infrastructure alone does not build social cohesion (Cortese 1982). Some degree of social cohesion seems to have formed in Williston and the 1970s and 1980s boomtown literature depiction of a frayed community did not emerge, but largely because people in Williston made it an active priority to protect and enhance a community that they loved.

Resource Extraction and Local Communities

The rich shale deposits, the technology, the global economy, and geopolitics all combined to make western North Dakota a center of the energy revolution. Prices will likely go back up—even if it isn't immediate, geopolitics will again shift, and gas and oil extraction will likely continue, despite the many larger and serious environmental impacts of fracking. When they do, the imperative that pushes for rapid profit will likely lead to the extraction of the resources at a pace and scale that will pose the challenges of previous booms. Under conditions of a boom, there is no reason for local and state governments to give tax breaks. Indeed, there should be a higher tax burden on extraction to defray costs of needed infrastructure, such as roads, bridges, and water and sewer lines. There should also be cost-sharing programs, and the companies, abiding by local zoning laws, should contribute to protecting the integrity of the local community. Further research in western North Dakota can unveil details of financing strategies, state regulations, and other local responses. Strategies employed in other municipalities or regions may likely provide more details on implementation.

In keeping with the warnings of boomtown literature of the 1970s, we might continue to ask, "How does a local community protect its autonomy from external economic forces?" We conclude from this chapter that good planning is paramount. While we might assume that, of course, cities and counties engage in comprehensive and participatory planning processes, many do not, even in the face of unexpected external forces. Córdova (1986) concludes that the community with a plan is in a better position for defending itself in the face of external forces. If planning is so important, do planners have the power and expertise needed to adequately plan? Hans Bleiker (1980) suggests that, "to do boomtown planning is to do community planning under very difficult conditions" (1980, 146). Is planning education delivering enough planners who can plan in modern-day boomtowns? Both the Williams County Comprehensive Plan 2013 (2012) and the Williston-Williams County Regional Comprehensive Plan (2017) are excellent—and implementable—models of good planning for local planners and planning students, demonstrating planning that is "rigorous, responsive and responsible" (Bleiker 1980, 156).

While Williston officials attended to issues of social cohesion, not all longtime residents welcomed the disruption to what had been their way of life. A continuing question is how do you preserve and enhance integral social and cultural relations of communities given what Bleiker describes as a "continuous, relentless, unabated" attack on the quality of life of residents (1980, 147)? The love for their community that was demonstrated by Williston and Williams County officials helped create the political will to do what they could to protect long-term quality of life, including efforts to ensure that the extractive industries share the burden of costs associated with extraction.

Critical issues remain, however. Are true costs and consequences being measured, including the geologic and environmental impacts from gas flares and the extensive use of chemicals and water? Williston and Williams County were forward thinking in their use of their powers—powers that are limited—but how much should local communities have to pay for the profits and policies favoring extractive industries without the authority or autonomy to regulate them? Depending on the price in the market, the extraction seems to occur at increasingly accelerated rates, with impacts to communities at a pace beyond what is reasonable to handle. If only the energy economy could be slowed, or shifted, it would allow for planning, environmental assessments, mitigation measures, and perhaps the development of better alternatives, admittedly easier said than done.

The world witnessed the efforts to call attention to the environmental impacts of extracting oil. In 2016 and 2017, Indigenous leaders at Standing Rock and surrounding communities, with the support of the Indigenous Environmental Justice Network (IEN), alerted the nation to the contamination of water sources that would follow from spills if the Dakota Access Pipeline, which would trans-port oil from the Bakkens, was built (#NoDAPL). While support for their effort was widespread, state and federal forces were called to squelch the protests, and many Indigenous protesters faced criminal charges, although most were later dropped. Critical to take note of was that North and South Dakota, using an ALEC (American Legislative Exchange Council) template, passed a law that made protest on land owned by energy companies a felony. Challenges to its constitutionality, however, may prevent the law's implementation. Despite the industry claims that the pipeline was secure and the leaks would not occur, several already have and in November 2019, as widely reported in the media, 380,000 gallons of oil spilled in an area three hundred miles east of Williston, seeping into the wetlands below.

Concluding Thoughts

When fracking came to western North Dakota, citizens and civic leaders of Williston, ND, remembered the boom and crashing bust of the 1980s. It wasn't going to happen that way again. This time, they were going to assert some con-trol and actively guide the effects of the accelerated economic and population growth. This preliminary case study makes the case that the worst possible civic response to a boomtown situation is to be so "grateful" for the attention and economic presence that doors are wide open to allow whatever conditions are sought by industry actors. Instead, civic leaders that aren't afraid to seek the long-term greater interests of their communities position themselves more favorably and increase the likelihood of reaping more of the positive benefits of this economic activity and fewer of the negative impacts. So when Williston leaders "came with a whole different knowledge of what they were going to do this time," it meant that they provided another model of how to respond to boomtown phenomena or more largely how, as a local community, to respond not only to actors external to their community, but actors in a powerful industry. This chapter is the story of planning in a modern-day boomtown— meaning that despite the power and dynamics of the global energy economy, civic leaders could play an active role in shaping how they would be affected using the power of local and state government. The modern-day boomtown is the one that stands up for itself and asserts some autonomy to protect its long-term interests—even in the face of the vulnerabilities from external forces. There is a potential that in the defense of local communities, many of the much larger issues may be forced to the surface.

Given the limitations from state government and the pressures of this powerful industry, to what extent can local communities have an impact on whether and under what conditions resource extraction occurs in their communities? Is it possible that they can do more than yield to demands placed upon them from external economic entities such as gas and oil companies? In this case, these particular external actors were themselves disrupting the order of the energy industry, affecting national energy policy and global geopolitics. Initially, it was not the big energy corporations, but "the independents" that bet on fracking. These company people may have felt that they did not have a choice but to comply with local demands and expectations—at least initially. Perhaps they just thought it would be easier or that it was the right thing to do. Is it also possible that they saw both the overall merits as well as the benefits to their industry if they found a way to comply and work with local and state entities that sought conditions that would not only mitigate many of the day-to-day impacts for longtime residents and newcomers, but assure long-term survival of the boom and bust cycles of extractive industries?[5]

The transportation of the black gold poses challenges as well. There are potential accidents from railcars carrying fuels and pipelines across sacred lands disrupting landscapes and water sources, leaving open the threat of oil spills and the contamination that follows. The concerns that Indigenous protesters raised in 2016 remain salient.

As the industry becomes more entrenched in the state, do industry actors start to exert more influence in state legislatures away from the interests of the state, its landowners, or its Indigenous and local communities? These questions are left for another day, or another researcher.

With this focus on civic response to the rapid influx of money and people, we were particularly interested in the relationship between local communities and the forces of an outside industry. Much of the extracted oil is exported. From many directions, there is a push for renewable energy sources, particularly as the projected path of global climate change becomes more ominous. If done at all, what would it take for the pace of shale production to be slower for more sustainable extraction? The focus of this chapter does not preclude an awareness or sensitivity to the broader debates about national energy policies: the environmental impacts of burning methane, injecting chemicals into the earth, using massive amounts of precious water, groundwater pollution, radioactive contamination, and earthquakes.

Horizontal drilling and hydraulic fracturing made it possible for the United States to now produce more crude oil than any other country, surpassing both Russia and Saudi Arabia, and to export more oil than it consumes (US Energy Information Administration 2018). But the United States still consumes a great deal of oil—at a rate far greater than any other country. The move away from fossil fuel energy toward both conservation and renewable sources seems urgent in light of the many reports from climate scientists firmly establishing the ticking clock.

While this chapter's focus does not include an exploration of these issues, as community development scholars, we should continue to think about the impacts on the social and cultural fabric of a community and its vulnerability to forces external to that community. A central question is the role of community and regional planning to preserve, protect, and develop local communities and think how government policies and practices best respond to the ebbs and flows of national and international trends in energy production and politics. The lessons of the chapter suggest that—at the very least—exchanging the right to extract resources from one's jurisdictional authority should be met with the responsibility to minimize and mitigate the negative impacts, while helping to increase the likelihood of the long-term viability of the communities that are the sites of this economy activity. Coming in, extracting, and then just walking away is not acceptable in the modern-day boomtown. Despite limitations, and in the face of forces outside its control, the case of Williston, North Dakota, demonstrates proactive efforts toward long-term community stability.

Notes

1. UNM students who traveled to western North Dakota in March 2012 included Weston Archuleta, Amy Rincón, Charles Joslin, Alvin Eugene Salazar, and Michaela Shirley. Assisting the travel team were Kiersten Mathsen and Grace C. Smith.

2. In 2013, USGS estimated that 7.4 billion barrels of oil, 6.7 trillion cubic feet of associated/dissolved natural gas, and 0.53 billion barrels of natural gas liquids exist in the Bakken and Three Forks Formations (US Department of the Interior 2013).

3. In 2015, the state legislature passed a law to reduce the percentage to 5 percent and by early 2019 new legislation was introduced to raise the rate back to 6.5 percent. The debates around the tax rate center around the amount of revenue that would be generated for infrastructure and education needs versus the argument a higher extraction tax discourages the industry from gas and oil extraction in the state.

4. In figures released in June 2019 by the North Dakota Department of Mineral Resources, North Dakota was producing 1.42 million barrels of oil per day and 2.88 million MCF per day. The number of rigs increased from the previous year and flaring of natural gas was up (Hughlet 2019).

5. Interviews with industry representatives that interface with the local government officials could add to further research on the issue.

References

Armstrong, Josh. 2013. "What the Frack Can We Do: Suggestions for Local Regulation of Hydraulic Fracturing in New Mexico." *Natural Resources Journal* 53:357–81.

Biermann, Sharon, Doina Olaru, and Valerià Paül. 2016. *Planning Boomtown and Beyond*. The Planning and Transport Research Centre (PATREC). Crawley, W.A.: University of Western Australia.

Bleiker, Hans. 1980. "Community Planning in Boomtowns: Why It's Not Working Very Well and How to Do It More Effectively." In *The Boomtown: Problems and Promises in the Energy Vortex*, edited by Joseph Davenport and Judith A. Davenport, 145–56. Laramie: University of Wyoming, Department of Social Work, Wyoming Human Services Project.

Brady, Jeff. 2018. "After Struggles, North Dakota Grows into Its Ongoing Oil Boom." National Public Radio, November 23. https://www.npr.org/2018/11/23/669198912/after-struggles-north-dakota-grows-into-its-ongoing-oil-boom.

Briody, Blaire. 2017. *The New Wild West: Black Gold, Fracking, and Life in a North Dakota Boomtown*. New York: St. Martin's Press.

Brown, Chip. 2013. "North Dakota Went Boom." *New York Times Magazine*, January 31.

Caraher, William, and Kyle Conway. 2016. *The Bakken Goes Boom: Oil and the Changing Geographies of Western North Dakota*. Grand Forks: Digital Press at the University of North Dakota.

Chambers, William. 1933. "Kilgore, Texas: An Oil Boom Town." *Economic Geography* 9 (1): 72–84. Accessed April 9, 2012. JSTOR Arts and Sciences II, EBSCOhost. https://doi.org/10.2307/140803.

City of Williston. 2017a. Overview and Recent History of Major Revenue Sources and Uses. September 7.

City of Williston. 2017b. Williston-Williams County Regional Comprehensive Plan: An Update to the 2010 Williston Comprehensive Plan and Transportation Plan. Adopted March 16, 2017. https://www.cityofwilliston.com/departments/planning_and_zoning/williston_williams_coutny_regional_plan_update.php.

CNN Business. 2015. "Oil Boomtown on the Brink?" January 22. Video. https://www.youtube.com/watch?v=r34Ul94kY6E.

Córdova, Teresa. 1986. "National Organizations and Local Communities: Land Use and Social Conflict in Southern Colorado." PhD diss., University of California, Berkeley.

Cortese, Charles F. 1979a. "Rapid Growth and Social Change in Western Communities." *Social Impact Assessment* 40 (41): 7–11.

Cortese, Charles F. 1979b. "Social Aspects of Community Impact." In *Energy Impacted Communities Management*, edited by Robert V. Johnston, David B. Longbrake, and John S. Hale, 37–62. Denver, CO: University of Denver.

Cortese, Charles F. 1979c. "Social Impacts of Energy Development in the West: Introduction." *Social Science Journal* 16 (2): 1–7.

Cortese, Charles F. 1982. "The Impacts of Rapid Growth on Local Organizations and Community Services." In *Coping with Rapid Growth in Rural Communities*, edited by Bruce A. Weber and Robert E. Howell, 115–35. Boulder, CO: Westview Press.

Cortese, Charles F., and Bernie Jones. 1977. "The Sociological Analysis of Boomtowns." *Western Sociological Review* 8 (1): 76–90.

Davies, Phil. 2012. "Desperately Seeking Workers in the Oil Patch." Federal Reserve Bank of Minneapolis. Fedgazette, April 18.

Dittrick, Paula. 2011. "Trucking the Bakken." Oil and Gas Journal, July 25. https://www.ogj.com/articles/print/volume-109/issue-30/regular-features/journally-speaking/trucking-the-bakken.html.

Donovan, Lauren. 2012. "Williams County Extends Man Camp Ban." *Bismarck Tribune*, March 3. bismarcktribune.com/news/state-and- regional/williams-county-extends-man-camp-ban/article_87024584-6505-11e1-8140-001871e3ce6c.html.

Fernando, Felix N., and Dennis R. Cooley. 2016a. "An Oil Boom's Effect on Quality of Life (QoL): Lessons from Western North Dakota." *Applied Research in Quality of Life* 11:1083–1115. https://doi.org/10.1007/s11482-015-9422-y.

Fernando, Felix N., and Dennis R. Cooley. 2016b. "Socioeconomic System of the Oil Boom and Rural Community Development in Western North Dakota." *Rural Sociology* 81 (3): 407–44.

Freudenburg, William R. 1981. "Women and Men in an Energy Boomtown: Adjustment, Alienation, and Adaptation." *Rural Sociology* 46 (2): 220–24.

Gilmore, John S., and Mary K. Duff. 1975. *Boom Town Growth Management: A Case Study of Rock Springs-Green River, Wyoming.* Boulder, CO: Westview Press.

Gogal, Danny. 2014. "Collaborative Problem Solving: A Tool to Address Fracking Concerns." *EPA Blog*, June 5. https://blog.epa.gov/2014/06/05/collaborative-problem-solving/.

Gold, Russell. 2014. *The Boom: How Fracking Ignited the American Energy Revolution and Changed the World.* New York: Simon and Schuster.

Gramling, Robert, and Sarah Brabant. 1986. "Boomtowns and Offshore Energy Impact Assessment: The Development of a Comprehensive Model." *Sociological Perspectives* 29 (2): 177–201.

Gulliford, Andrew. 1989. *Boomtown Blues: Colorado Oil Shale, 1885–1985.* Boulder: University of Colorado.

Healy, Jack. 2016. "Built Up by Oil Boom, North Dakota Now Has an Emptier Feeling." *New York Times*, February 7.

Hodur, Nancy M., and Dean A. Bangsund. 2013. "Population Estimates for City of Williston." Agribusiness and Applied Economic Report 707. North Dakota State University, August.

Hughlett, Mike. 2019. "North Dakota Posts Record Gas and Oil Production Numbers." *Star Tribune*, August 15.

Karl, Terry. 2007. "Oil-Led Development: Social, Political, and Economic Consequences." CDDRL Working Papers. Center on Democracy, Development, and the Rule of Law. Freeman Spogli Institute for International Studies. Stanford University, January.

Kohrs, ElDean V. 1974. "Social Consequences of Boom Growth in Wyoming, Casper, WY: Central Wyoming Counseling Center." Paper presented at the Rocky Mountain American Association of the Advancement of Science Meeting, Laramie, Wyoming, April 24–26.

Jacquet, Jeffrey B., and David L. Kay. 2014. "The Unconventional Boomtown: Updating the Impact Model to Fit New Spatial and Temporal Scales." *Journal of Rural and Community Development* 9 (1): 1–23.

Jones, Bernie, and Charles Cortese. 1976. "Patterns of Boom Town Experiences, Implications for Future Work in the Field of Social Impact Assessment." In *Annual Meeting of the Society for Social Problems*, 27–30. New York, August.

Little, Ron. 1977. "Some Social Consequences of Boomtown." *North Dakota Law Review* 52 (3): 401–25.

Madrick, Jeff. 2013. "The Anti-Economist: A Bit of Good News." *Harper's Magazine*, April 2013, 12. Print.

Mahon, Joe. 2005. "Black Gold Rush." Federal Reserve Bank of Minneapolis. Fedgazette, September 1.

Manning, Richard. 2013. "Bakken Business: The Price of North Dakota's Fracking Boom." *Harper's Magazine*, March 2013, 34.

Manning, Robert. 2014. *The Shale Revolution and the New Geopolitics of Energy*. Washington, DC: Atlantic Council.

Markusen, Ann R. 1978. *Socioeconomic Impact Models for Boomtown Planning and Policy Evaluation*. Institute of Urban and Regional Development. Working Paper 285. Berkeley: University of California.

Melberg, Mitch. 2019. "Williston Leads State in Taxable Sales and Purchases Increases for 2018." Williston Herald, May 1.

Neighborhood Scout. 2018. "Williston Crime Analytics." September. https://www.neighborhoodscout.com/nd/williston/crime.

Nicas, Jack. 2012. "Oil Fuels Population Boom in North Dakota City." *Wall Street Journal*, April 6. https://www.wsj.com/articles/SB10001424052702304072004577328100938723454.

Nolon, John R., and Jessica A. Bacher. 2014. "Mitigating the Adverse Impacts of Hydraulic Fracturing: A Role for Local Zoning?" Pace Law Faculty Publications. Pace University. https://digitalcommons.pace.edu/lawfaculty/957/.

Northwest Landowners Association. 2019. Media Advisory, July 29, 2019.

Oyakawa, Javier. 2014. "Economic Impacts of the Eagle Ford Shale: Modeling and Data Issues." In Mid-Continent Regional Science Association 45th Annual Conference Proceedings, Madison, Wisconsin, June 3–5, compiled by John Leatherman, 58–75. http://www.mcrsa.org/wp-content/uploads/2018/03/2014_Proceedings-1-1.pdf#page=66.

Piccolo. 2008. "The Bakken Formation: How Much Will It Help?" The Oil Drum: Discussions about Energy and Our Future, April 26. http://www.theoildrum.com/node/3868.

Pitman, Janet, Leigh C. Price, and Julia A. LeFever. 2001. "Diagenesis and Fracture Development in the Bakken Formation, Williston Basin: Implications for Reservoir Quality in the Middle Member." US Department of the Interior. US Geological Survey Professional Paper 1653.

Population.us. 2014. Population of Williston, ND. https://population.us/nd/williston/.

Prairie Public Broadcasting. 2012. "Faces of the Oil Patch," May 17. https://www.youtube.com/watch?v=5GWMpEnK3Vg.

Russell, Karl. 2016. "How Oil Prices Are Falling Again, Explained in Four Charts." *New York Times*, July 29. https://www.nytimes.com/interactive/2016/business/energy-environment/oil-price-supply-demand-imblance.html.

Schulz, Bailey. 2018. "Oil Bust Was Actually Good for a Town Called 'Shale City.'" BloombergQuint, March 12. https://www.bloombergquint.com/business/the-oil-bust-was-actually-good-for-shale-city-north-dakota#gs.l0ixtJu1.

Shigley, Paul. 2009. "When the Boom Busts." *Planning*, April, 18–21.

Smith, Michael D., Richard S. Krannich, and Lori M. Hunter. 2001. "Growth, Decline, Stability, and Disruption: A Longitudinal Analysis of Social Well-Being in Four Western Rural Communities." *Rural Sociology* 66 (3): 425–50. https://doi.org/10.1111/j.1549-0831.2001.tb00075.x.

Sontag, Deborah, and Robert Gebeloff. 2014. "The Downside of the Boom: North Dakota Took On the Oversight of a Multibillion-Dollar Oil Industry with a Regulatory System Built on Trust, Warnings, and Second Chances." *New York Times*, November 22. https://www.nytimes.com/interactive/2014/11/23/us/north-dakota-oil-boom-downside.html?_r=0.

Springer, Patrick. 2018. "North Dakota 'Just Won the Geology Lottery': Oil Exec Estimates the Bakken Reserves Hold 30 to 40 Billion Reserves of Recoverable Oil." Inforum: Powered by the Forum and WDAY, September 24. https://www.inforum.com/business/energy-and-mining/4503653-north-dakota-just-won-geology-lottery-oil-exec-estimates-bakken.

Summers, Gene F., and Kristi Branch. 1984. "Economic Development and Community Social Change." *Annual Review of Sociology* 10:141–66. https://doi.org/10.1146/annurev.so.10.080184.001041.

US Census Bureau. 2000. *Decennial Census*. ndsu.edu/sdc/publications/census/NDcities1920to2000.pdf.

US Census Bureau. 2017. "Quick Facts: Williston City, North Dakota; Williams County, North Dakota." https://www.census.gov/quickfacts/fact/table/willistoncitynorthdakota,williamscountynorthdakota,nd/PST045217.

US Department of the Interior. 2013. "USGS Releases New Oil and Gas Assessment for Bakken and Three Forks Formations: Finds Formations Have Greater Resource Potential than Previously Thought." Press Release, April 13. https://www.doi.gov/news/pressreleases/usgs-releases-new-oil-and-gas-assessment-for-bakken-and-three-forks-formations.

US Energy Information Administration. 2018. "The United States Is Now the Largest Global Crude Oil Producer." September 12. https://www.eia.gov/todayinenergy/detail.php?id=37053.

US Geological Survey. n.d. "How Much Oil and Gas Are Actually in the Bakken Formation?" https://www.usgs.gov/faqs/how-much-oil-and-gas-are-actually-bakken-formation?qt-news_science_products=0#qt-news_science_products.

Valentine, Katie. 2015. "Crime in North Dakota's Oil Boom Towns Is So Bad That the FBI Is Stepping In." Think Progress, March 6. Accessed March 10, 2019. https://thinkprogress.org/crime-in-north-dakotas-oil-boom-towns-is-so-bad-that-the-fbi-is-stepping-in-76e3203eab24/.

Veeder, Gene. 2012. In "North Dakota's Oil Rush: If the Boom Goes Bust." Wall Street Journal. WSJ Live, December 23. Video. https://www.youtube.com/watch?v=Nm-ESEfxpgU.

Weber, Bret, Julia Geigle, and Carenlee Barkdull. 2014. "Rural North Dakota's Oil Boom and Its Impact on Social Services." *Social Work* 59 (1): 62–72.

Weller, R. 2009. *Boomtown 2050: Scenarios for a Rapidly Growing City*. Crawley, W.A.: University of Western Australia Press.

Williams County. 2012. Comprehensive Plan 2035. Williams County, North Dakota. Adopted December 20, 2017.

Yardley, William. 2016."In North Dakota, an Oil Boomtown Doesn't Want to Go Bust." *Los Angeles Times*, January 1. https://www.latimes.com/nation/la-na-sej-north-dakota-oil-town-20160111-story.html.

LOCAL PLANNING IN BEAVER COUNTY AND THE SHELL CRACKER PLANT

Sabina E. Deitrick and Rebecca Matsco

The Marcellus Shale industry has developed and expanded greatly in Western Pennsylvania in little more than a decade, and no single project has taken on the size and extent of the Royal Dutch Shell ethane cracker plant in Potter Township, Beaver County. The cracker plant, which will begin operations in 2022 (Gough 2021), marks a major expansion of the natural gas industry into downstream supply chain production, something Pennsylvania has sought since natural gas extraction in the Marcellus Shale boomed late in the first decade of this century. The Marcellus and Utica Shales of Western Pennsylvania, Ohio, and West Virginia are rich in natural gas liquids (NGLs) used to produce ethylene for the supply of plastic products (IHS Markit 2017). The ethane cracker plant marks the first facility of this kind in the Marcellus Shale region and the first such investment by Royal Dutch Shell outside the Gulf Coast area in twenty years (Davis 2016).

The location of the facility, Potter Township, is a small municipality in Beaver County in Western Pennsylvania. The ethane cracker plant project lies in a contested space between state economic development initiatives and local governance and planning, if not in a more contested space between environmental activism and progrowth economic development policy. The roles of public administration and planning capacity became important in understanding the policies and plans to bring the project forward into the public realm and the state's role in promoting the development. The cracker plant initiative shows how Pennsylvania continues to challenge local-level cooperative planning by forcing local officials into reactive positions under state directives, with limited capacity for cross-municipal engagement. However, in learning how to expand its policy and planning capacity, Potter Township represents a more complicated scenario where collaboration and community engagement became the means for the township to develop an active planning and public administration position, from initially being forced into reactive positions by the state.[1]

Background

Potter Township faced a double economic whammy in less than a year: in the fall of 2011, Horsehead Corporation, owner and operator of a zinc smelting facility in Potter Township, announced that it would close its Potter Township plant when it opened its new state-of-the-art facility in North Carolina in 2013. Potter Township would not only lose a major employer in the region, with over six hundred jobs, but stood to face the loss of more than one-fifth of its $700,000 annual budget generated by property taxes collected from Horsehead and earned income taxes collected from its workforce. Furthering the blow was the prospect of a 300-acre brownfield (Stonesifer 2018b). The Horsehead site was once the largest zinc smelter plant in the country and contained land contaminated by lead, arsenic, cadmium, and mercury left by Horsehead over nine decades (Litvak 2015).

Not long after, in March 2012, then governor Tom Corbett of Pennsylvania signed legislation that gave Royal Dutch Shell $1.7 billion in tax credits for twenty years to build and operate an ethane cracker facility in Pennsylvania. Pennsylvania officials' preferred location was Potter Township, a small community of about five hundred residents in Beaver County, but, more importantly, the location of the Horsehead Corporation's zinc smelting plant where Shell Chemicals, a subsidiary of Shell Oil, had secured an option on the site (Wereschagin and Vidonic 2012). The prospects for the new investment were celebrated by state politicians in two places—in a nearby community, Hopewell Township, on March 12 and at a press event on March 16, with then governor Corbett making the announcement thirty miles away in downtown Pittsburgh. Neither took place in Potter Township nor featured its officials, though they were invited to attend the surprise event.

This left Potter Township in a situation of many unknowns. Significant parts of both the Marcellus and Utica shale plays contain natural gas liquids (NGLs), high-value ethane and propane that provide the raw materials for petrochemical production. Shell was attracted to the "wet gas" of the Marcellus and Utica shale formations that can be separated in a cracking facility, where ethane is converted to ethylene and, ultimately, polyethylene to become feed-stock for various plastics products.

Shell had also been considering sites in nearby Ohio and West Virginia for the new facility outside its long presence on the Gulf Coast, but Horsehead's announcement in 2011 that it would close its Potter Township facility at the end of 2013 enhanced the prospects of Royal Dutch Shell selecting Pennsylvania for its facility. Governor Corbett's tax break of $1.7 billion came under the state's Keystone Opportunity Expansion Zone program, with additional state and local tax breaks for the international firm (Boselovic 2016; Bumsted 2012). In anticipation of the project, the Pennsylvania legislature exempted oil and gas land use from local controls, further reducing the power of the township to plan with the firm (see chapters 2 and 3).

For Potter Township, this new investment backed by tax abatements and new limitations on local land use planning was not going to translate into revenue for the township's budget. With such a sizable investment in such a small community—an investment sure to pit opponents from the environmental side against proponents from the economic development side—could Potter Township address all the challenges that would be coming its way? Did Potter Township have the capacity to act or, as a small government, be only reactive against the much larger forces of both state subsidies and a global firm? Would its capacity in local administration and land use planning be further constrained by anti-Shell opponents on the environmental side? This chapter addresses the capacity and capabilities of small local governments in planning and public administration on issues related to the shale gas boom in the region. By analyzing Potter Township's processes after the Shell announcement, this chapter argues that small municipal governments can develop capacity to meet extensive planning and public administration challenges in the wake of the natural gas boom in the Marcellus Shale. While unable to turn the tide of the state's economic development growth coalition or reflect stronger environmental stands advocated by activists, Potter Township resolved important planning and public administration areas to gain benefits it would not have received under the direction of the state.

The Role of the State (of Pennsylvania)

Pennsylvania has long engaged in tax incentive programs but expanded its development incentives in the latter part of the twentieth century. The consequences of rapid deindustrialization and manufacturing job loss in the late 1970s and 1980s created concentrated areas of disinvestment and unemployment (Bluestone and Harrison 1983). As communities faced fiscal distress from plant closings and job losses, states began to design tax incentive programs that were tied to specific locations, and new "enterprise zones" spread across older industrial regions in the United States and overseas (Wolf 1990).

In 1998, Pennsylvania began its Keystone Opportunity Zone (KOZ) program aimed to revitalize economically distressed communities by attracting new business. KOZs were meant to encourage investment in designated distressed areas of the state, with firms receiving 100 percent tax abatement of real estate property taxes from the three taxing bodies in the state—municipality, school district, and county (Holoviak and Carabello 2008). Based on the Michigan Renaissance program, Pennsylvania's KOZs were originally designed as a "one-time only" program (Pennsylvania Legislative Budget and Finance Committee 2009). The tax abatement aimed to bring new businesses to distressed properties and represented, in the view of the state, an investment in future growth. Despite Pennsylvania's claim that the KOZ program was "unique" (Pennsylvania Department of Community and Economic Development 2015, 3), by 1998, at least thirteen states had some form of an enterprise zone (EZ) program (Peters and Fisher 2004).

The KOZ program was extended through the Keystone Opportunity Expansion Zone (KOEZ). To be established, a new KOEZ selected by a local community needs to meet at least two of twelve criteria demonstrating distress and be approved by the two state bodies overseeing the program, the Pennsylvania Department of Community and Economic Development (DCED) and the Pennsylvania Department of Revenue and Department of Labor and Industry. Despite continuing the "one-time" KOZ program into the KOEZ expansion, there has been little estimation of the impacts of Pennsylvania's enterprise zones on job creation and municipal revenues or consideration of any possible negative effects. Holoviak and Carabello (2008) evaluated Pennsylvania's KOZ/KOEZ and gave mixed results on their effects in rural regions of the state. Even adding qualitative studies of the positive impacts demonstrated in select cases, their final analysis concluded that "the KOZ designation is the marginal deciding factor in business location and expansion" (Holoviak and Carabello 2008, 16).

Academic research and professional evaluations have long suggested that enterprise zone programs are largely unsuccessful, the benefits overstated, and the tax incentives not favorable in affecting business location decisions (Greenbaum and Landers 2009; Peters and Fisher 2004). However, to understand better why they do get implemented, Greenbaum and Landers (2009, 472) reflected:

> There may be a political component to ignoring the EZ literature, as the literature may not jibe with policy makers' perceptions of constituent

preferences or their perceptions of program outcomes based on information they receive from constituent groups such as district voters, development officials, or business, professional, or trade groups.

Pennsylvania policy makers fit that description. State officials, with little expertise in the new gas industry and its related industrial activities, promoted tax abatement policies familiar to government officials. Despite marginal evaluations of success, Pennsylvania officials pursued the KOEZ as their means to attract Shell, with tax credit benefits worth $66 million per year for twenty-five years (Carrington and Davies 2015).

The KOEZ in Potter Township stands out in additional ways in how the state promoted the program. The state imposed the tax abatement on Potter Township without its initial consideration or input. Not only was there no local "empowerment" in this process as a "unique state/local partnership," as state literature described the program (Pennsylvania Legislative Budget and Finance Committee 2009), there was no local engagement in establishing the KOEZ, which ran counter to state law (Samuel 2012). As local officials were caught off guard and learned that their community was to become an expanded KOEZ for a Shell cracker plant at the governor's press conference, a news summary of the state's announcement reported that "nobody thought to ask Potter Township about the KOEZ" (Stonesifer 2018b).

Further hampering Potter Township was its need to find new revenue sources to cover the losses from Horsehead's closure, and the state was doing little to find replacement. "We were told that Potter is a small player in a bigger scheme," Township Supervisor Rebecca Matsco said, adding that the supervisors were under "immense pressure" to sign off on the KOEZ despite their objections. "All we were asking was that Potter be recognized and compensated for our losses. There is a community here that matters" (Stonesifer 2018b).

Capacity and Capabilities

Potter Township faced difficult choices under many constraints imposed by the state. Planning and public administration, the two intertwined jobs of the three-person board of supervisors, are often linked to capacity in both the public administration and planning fields, with capacity largely defined by the available professional and budgetary resources (Loh 2015). By standard measures of public administration, Potter Township would be limited in capacity on multiple fronts to deal with the loss of revenue for its budget and limited expertise and resources available to engage in the necessary planning the Shell plant would entail.

While the planning literature is not as coherent on measures of capacity as the public administration field, by most measures, capacity in planning and public administration for local governments is often defined by (a) inputs—resources available to professionals to carry out their positions and make decisions—and (b) outputs—the capabilities—or abilities—that determine actions toward tasks or goals (Loh 2015).

Most definitions of capacity focus on availability of resources, or inputs. By traditional public administration measures of capacity as resources and professional expertise, Potter Township—a second-class township with a volunteer set of three supervisors and an annual township budget of $700,000—faced capacity issues. When defined by available inputs, capacity is certainly minimal or constrained for second-class townships in Pennsylvania (Loh 2015; Sokolow 1981).

Honadle (2001), in reflecting on the ability of local governments to take on new responsibility coming from federal and state governments, finds that

> whether a local government has capacity to assume new responsibilities depends on how the existing capacity is being used. Capacity is a relative term, something that only makes sense compared to other places with more or less capacity or to a time in the past or in the future when there was or there is expected to be more or less capacity. It does not make sense logically to take "stock" of all of the endowments of a local government at a point in time and, from that, arrive at a sum equal to that governmental unit's capacity to handle new responsibilities. (Honadle 2001, 7)

Potter Township's capacity to handle the new Shell development would be tested. With a three-person supervisory board accountable for the township's budget of revenues—derived mainly from property taxes and dominated by expenditures for fire and other public safety services—by public administration and planning capacity, Potter Township would be challenged and reactive to the position the state put them in with the KOEZ tax holiday offered to Royal Dutch Shell. The challenges included the following:

1. Reduction of property taxes collected on the site, from over $40,000 by Horsehead to a projected zero under the KOEZ tax break, in addition to employment taxes, for a total loss of $100,000.
2. Need for assessment of the site for environmental conditions. Horsehead was cited repeatedly by Pennsylvania Department of Environmental Protection (DEP) and the transfer of the site from Horsehead to Shell

required a legal agreement on Act 2, Pennsylvania's voluntary land recycling program (Litvak 2015).

3. Opposition to the cracker plant would come from outside the township,
   with nearly all local residents supporting the new investment by Shell and
   well-organized environmental groups against it.

With limited capacity, as measured by resources, or inputs, would Potter
Township be able to construct a route for local government to meet the planning
and public administration challenges they would be facing with the Shell cracker
facility? As Honadle argues, capacity is more than the resources definition, and
for Potter Township: "The capacity needs of local governments will depend on
the programmatic challenges they are dealt by the policymakers in Washington
and state capitals" (Honadle 2001, 12). An inventory of capacity at that point
would not have been predictive of the township's capacity to learn and build from
the Shell deal imposed by the state.

We find that the case of Potter Township falls into the scope of small governments overcoming limitations imposed by traditional resource measures of
public administration and planning capacity to deal with the challenges faced. To
conceptualize this, it involves expansion of the notion of capacity to encompass
other critical resources, including expertise and community engagement. With
additional community engagement and resources, we can analyze Potter Township's process through the Shell project in different and more conceptual ways.
Understanding capacity expands beyond inputs of financial resources, the typical constraints for small municipalities. This expanded notion of capacity now
includes the notion of commitment (Norton 2005). Commitment, or "the will
to act" (Loh 2015, 136), extends the processes in planning and public administration beyond capacity defined by resources to capabilities defined by linking
resources to action, with action achieved through community engagement.

The idea and conceptualization of commitment in the discussion and analysis
of capacity in local government becomes central in the Potter Township case.
What the township lacked in traditional capacity and the inputs, or resources,
to undertake the Shell cracker plant project with its public administration and
planning challenges, it countered with commitment (Stonesifer 2016). The township advanced its agenda and resisted reactive planning through commitment to
bringing back revenues, providing services for residents, and gaining additional
resources from the Shell project the state imposed on the community. The will to
act became engrained through a community engagement process that overcame
the lack of capacity as understood by traditional planning and public administration measures.

TABLE 7.1 Major events of Shell Cracker Plant announcement and process, Potter Township, PA

2011	Horsehead Corporation announces it is building a new zinc facility in North Carolina and will be closing its Potter Township, Beaver County, plant.
2012, March	Shell announces it is looking at the Horsehead site to build an ethane cracker plant in Potter Township, Beaver County, PA.
2012, March	Governor Tom Corbett announces that Shell will receive tax credits in the form of a Keystone Opportunity Expansion Zone.
2012	Shell buys nearby properties for $12 million.
2014	Route 18 Corridor Analysis Report completed for the Beaver County Corp. for Economic Development.
2015	Pennsylvania Department of Environmental Protection (DEP) releases details of Shell's pipeline project.
2016	Shell installs LED lighting at the site to reduce glare and visibility.
2017	Construction officially starts November 8.
2018	Shell donates $1 million to Community College of Beaver County's process technology program.

Capacity and Community Engagement

Before any notice of the Shell cracker plant, Potter Township was typical of other Western Pennsylvania municipalities, experiencing long-term population decline, with aging population and infrastructure. As required, Potter Township had a comprehensive plan on record, but little had been done to make the plan a reality, and the zoning ordinance dated from 1972. The township was familiar with the energy sector, but in different ways from the Shell plant. The first full-scale commercial nuclear power plant in the United States was four miles away at Shippingport, with a coal-fired power plant next door. For a community steeped in industrial manufacturing workers and chemical plants, cleaning up the contaminated Horsehead site was a positive for many in the community.

On the public administration side, the municipality was not fiscally distressed, as KOEZ communities are, and already engaged in many cost-saving, shared-service ventures with nearby municipalities, including road maintenance, public safety, purchasing, and auditing. In the minds of local officials and residents, the KOEZ programmed to relieve local distress was actually about to create it (Matsco 2018).

The process of community engagement became the means for the township to gain access to Shell and additional resources, while stemming the negative consequences of the already-decided KOEZ and its revenue implications for the township. Community engagement strategies became the prime means for the

Board of Supervisors to allow residents a space in the conflicted process of Shell and Potter Township. The Commonwealth of Pennsylvania convened the process, to some extent, through its involvement on the front end by setting up the tax holidays for the company that the local municipality could not change. After that, the situation was left to local officials.

The capacity issue could certainly be a major concern for many, under resource and input definitions, but expanding the conceptualization of capacity through the role of commitment can bring intangible resources to the community development process; and in community development, this includes relationships, partnerships, and political expertise.

Understanding capacity through community development also relates to civic capacity (Loh 2015). Here capacity expands through community embeddedness in the planning process and expands capabilities or outcomes. This was the case for Potter Township. Capacity building in the community came from previous planning and visioning efforts and learning to ask the "right" questions, in terms of community needs, and developing community expertise to assist in specific administration and planning decisions. Whatever Potter Township lacked in traditional input and output measures—resources and capabilities—intangible assets became the capacity-building components to try to direct resources their way in future dealings with the Shell ethane cracker.

Environmental

One of the biggest gains for the township was the Route 18 Corridor Analysis Study (KU Resources 2014). This study provided potential revenue sources and development opportunities within the township that had not been available for elected officials before. The study examined the road conditions and expansion options with the construction of the Shell plant and revenue possibilities for the township. The ability to advocate for and gain this analysis was critical for township supervisors in making subsequent planning decisions. The ultimate reconstruction of the road proved to be a gain for residents of the township (Stonesifer 2017).

Horsehead faced environmental challenges in its years in Beaver County, with a 2011 US Environmental Protection Agency (EPA) report finding zinc residue in the air and soil, and lead and cadmium leached from storage tanks. Shell proposed to the Pennsylvania Department of Environmental Protection (DEP) to cap the site, allowable under Act 2 Land Recycling Program, which challenged the federal Superfund law to allow clean-ups on brownfield sites to match new uses and reduce liability risks (Deitrick and Farber 2005). The cost was estimated at $89 million to build a five- to seven-foot cap of clean soil and to stabilize water through other devices, including retention ponds and a water treatment facility (Litvak 2015).

Many concerns about water were not under the direct jurisdiction of municipal government (Stonesifer 2018a). The township relied on state and federal agencies, academic researchers, and activists in the environmental community to provide relevant knowledge for the permitting process and helpful information for the public.

Planning and Public Administration

Potter Township officials were aware of the role they would play as the first of several likely cracker projects took shape in the Ohio River Valley. Some Beaver County residents and policy makers recalled the heavy impact of construction jobs when the Shippingport Atomic Power Station was built less than five miles away in the mid-1950s, but with no project of that scale in the decades since, there was no institutional memory from which Potter Township could operate. There was also virtually no assistance forthcoming from the state or federal government to help Potter prepare for a project of the cracker's scope. After Potter Township secured the Route 18 Corridor Study as a condition of approval for the KOEZ, the municipality began to adopt its recommendations. The main points were to completely overhaul local land-use policy with fresh zoning and new, more applicable ordinances. This allowed the township to establish a clear and predictable permitting process for the Shell project that would include adequate protections for the residential community, allow municipal officials to gain key technical information, and make room for the competing opinions and voices that would emerge from environmental and social justice advocates.

Always "top of mind" was the potential for Shell to determine that this cracker was no longer in its best interest and to close down the project after the site was remediated and the road rebuilt. This "top of mind" created conditions for commitment and the will to act for the township. It concentrated on becoming the community it had envisioned through its 2006 comprehensive plan, with or without the Shell facility. Based on a number of factors, including new zoning, the Route 18 study, and recommendations from the 2006 comprehensive plan, Potter Township moved to capture the rural nature of its surroundings by conserving a large swath of the Raccoon Creek Watershed that ran through the middle of the township before it could be infringed upon by unwanted development. These formerly industrially zoned "wild lands" were committed to uses no more intensive than residential and were further supported by conservation planning principles. A first-of-its-kind Natural Heritage District became a natural buffer between the new and old industrial activity in Potter Township and the quiet residential neighborhoods that the community desired. With its rich WWII history and enticing streambank, the Raccoon Creek Greenway is now the focal point of Potter's effort to build recreational tourism into its economic environment.

When measuring capacity by resources, Potter Township initially appeared to fall short on resources devoted to planning. Local public administrators were, however, able to expand capacity through intangible resources and collaborative action. These intangible resources are conducted through collaborations around planning and economic development between counties and smaller units of government (Loh 2015; Lobao and Kraybill 2009). Productive, cross-municipal collaborations meant that these "intangible relationships" (Loh 2015, 140) by township officials with others in local government and the professional community would make for more effective community planning.

Managed correctly, the public process of governmental decision-making is one of reconciliation that results in public policy that is equitable for diverse, often-competing viewpoints or constituent groups. On one side were residents who wanted no changes in the community, and on the other were those who thought the Shell plant would solve all municipal problems, however defined.

There were many other views. By the time of the final conditional use hearing in December 2016, protesters were arriving on buses to voice their concerns, and township supervisors were more aware than ever of the precedent Potter would set, not only for the Pittsburgh region but also for communities downstream that hoped for a project of their own. Potter Township residents offered a hospitable welcome to outside groups, and every attendee who wanted to speak was given time to do so. The hearing lasted thirteen hours over two days. Throughout the permitting phase, supervisors took care to consistently express the vision and values of the community. Officials were convinced that this was not an "either-or" choice between development versus conservation, but an opportunity for "all of the above" or "both-and" to grow an economically beneficial industry while preserving quality of life.

Potter's board had prepared for the aspects of development that were within the township's jurisdiction and of greatest concern to residents: light, sound, and traffic. The municipality was already subject to deceptive behavior and shoddy construction on a pipeline project in the community. Potter Township demonstrated a willingness to fight the industry by joining Robinson Township in its battle for local control and had argued in the Pennsylvania Commonwealth Court for—and won—the right to bond for seismic testing contractors working in the township. The expectations of the township were clearly stated, and, by the time of the required public meeting, Shell's site plan included downward-facing LED lighting, noise restrictions during construction, and a fully functional traffic management plan, coordinated with the local police department. Since the conditional use permit was issued, Potter successfully negotiated with Shell for the best possible pipeline route through the township to the plant that would affect the fewest property owners and minimize road openings and environmental impacts in the Raccoon Creek Greenway.

Transparency on the part of township officials enabled frank conversations about Potter Township's vision of its future. The result was a new public-private partnership for the Raccoon Creek Greenway for management and park development among the township, Shell, other industrial neighbors, and Beaver County's only land conservation group, the Independence Conservancy.

Capacity and Commitment Going Forward

The response to fracking and the Shell cracker plant in Potter Township shows how communities can expand capacity when they have capabilities of practice that link commitment and action. From its initially reactive position, Potter Township carved out a path to generate gains for the municipality and its residents when the state conceived of little involvement for the community in the construction of the Shell ethane facility. The "will to act" means that local governments have the ability to expand their capacity and meet formidable challenges that arise in municipal planning and public administration (Honadle 2001, 12):

> In considering local capacity, it is important to keep in mind that it is not a static concept. As knowledge and technology change, so does a unit of government's capacity to deal with problems. Likewise, as problems become more complex, more capacity may be needed to deal with them. It is also possible that the gap between the capacity a local government already has and the capacity it needs to manage certain types of programs is bridgeable through experience, technical assistance, the transfer of financial resources to the local government, or some other means of shoring up weak capacity.

Potter Township benefited from a number of capacity-building initiatives that expanded their ability to develop proactive, sustainable planning goals and meet the challenges such a large corporation and industrial development can bring. While part of the planning literature reflects a narrow view of capacity through resources or inputs available in solely monetary terms, the concept of community engagement brings a more extensive, nuanced view of capacity in the context of commitment, as expressed by the relationship between resources and action.

Many energy-dependent communities have strived to achieve sustainable development through changes in planning practices. Though these small communities often lack resources and capacity for complex planning approaches, many have engaged in community sustainability planning, "to develop strategies to leverage benefits while reducing losses resulting from boom-bust cycle" (Bowen, Civittolo, and Romich 2016). These strategies have benefited Potter Township. In boom-bust energy-dependent communities, expanding planning and public administration capacity through a dedicated role of community engagement can deliver more sustainable community planning.

Note

1. Chapter 2 analyzes the key Pennsylvania law changes that affected local-level planning controls.

References

Bluestone, Barry, and Bennett Harrison. 1983. *The Deindustrialization of America: Plant Closings, Community Abandonment, and the Dismantling of Basic Industry.* New York: Basic.

Boselovic, Len. 2016. "Good Jobs First Exec Sheds More Light on Corporate Subsidies." *Pittsburgh Post-Gazette*, September 17. https://www.post-gazette.com/business/development/2016/09/17/Good-Jobs-First-executive-director-Greg-LeRoy-talks-corporate-subsidies-at-Heinz-College-forum/stories/201609170045.

Bowen, Nancy, David Civittolo, and Eric Romich. 2016. "Community Planning Strategies for Energy Boomtowns." College of Food, Agricultural, and Environmental Sciences. Ohio State University Extension, June 23. https://ohioline.osu.edu/factsheet/cdfs-sed-6.

Bumsted, Brad. 2012. "Corbett's $1.7B Tax Pledge Banks on Jobs." TribLive Total Media, June 9. https://triblive.com/state/1944536-74/tax-million-corbett-deal-credit-pennsylvania-plant-shell-breaks-county.

Carrington, Damian, and Harry Davies. 2015. "US Taxpayers Subsidizing World's Biggest Fossil Fuel Companies." *Guardian*, May 12. Reprinted on Good Jobs First. https://www.goodjobsfirst.org/news/us-taxpayers-subsidising-worlds-biggest-fossil-fuel-companies.

Davis, Carolyn. 2016. "Pennsylvania Cracker Plant Underpins Shell's 'Cash Engine' Focus, Says CEO." Natural Gas Intelligence, June 8. https://www.naturalgasintel.com/pennsylvania-cracker-underpins-shells-cash-engine-focus-says-ceo/.

Deitrick, Sabina E., and Stephen C. Farber. 2005. "Citizen Reaction to Brownfields Redevelopment." In *Revitalizing the City: Strategies to Contain Sprawl and Revive the Core*, edited by Fritz W. Wagner, Timothy E. Joder, Anthony J. Mumphrey Jr., Krishna M. Akundi, and Alan F. J. Artibise, chapter 8. Armonk, NY: M. E. Sharpe.

Frazier, Reid. 2012. "Frequently Asked Questions about Ethane Crackers." Allegheny Front, March 15. https://www.alleghenyfront.org/frequently-asked-questions-about-ethane-crackers/.

Gough, Paul J. 2021. "Here's When Shell's Long-Anticipated Cracker Plant in Beaver County Will Be Operational." *Pittsburgh Business Times*, March 16. https://www.bizjournals.com/pittsburgh/news/2021/03/16/shell-plastics-plant-beaver-county.html.

Greenbaum, Robert T., and Jim Landers. 2009. "Why Are State Policy Makers Still Proponents of Enterprise Zones? What Explains Their Action in the Face of a Preponderance of the Research?" *International Regional Science Review* 12 (4): 466–79.

Holoviak, Paula A., and Damian Carabello. 2008. "An Evaluation of the Keystone Opportunity Zone (KOZ) and Keystone Opportunity Expansion Zone (KOEZ) Programs in Rural Pennsylvania." Harrisburg, PA: Center for Rural Pennsylvania, July. https://www.rural.palegislature.us/documents/reports/KOZ2008.pdf.

Honadle, Beth Walter. 2001. "Theoretical and Practical Issues of Local Government Capacity in an Era of Devolution." Journal of Regional Analysis and Policy 31:1–14. https://ideas.repec.org/a/ags/jrapmc/132195.html.

IHS Markit. 2017. "Prospects to Enhance Pennsylvania's Opportunities in Petrochemical Manufacturing." IHS Markit, March. https://teampa.com/wp-content/uploads/2017/03/Prospects_to_Enhance_PAs_Opportunities_in_Petrochemical_Mfng_Report_21March2017.pdf.

Kelsey, Timothy W., Mark D. Partridge, and Nancy E. White. 2016. "Unconventional Gas and Oil Development in the United States: Economic Experience and Policy Issues." *Applied Economic Perspectives and Policy* 38 (2): 191–214. https://doi.org/10.1093/aepp/ppw005.

KU Resources. 2014. "Route 18 Corridor Analysis Report." Beaver, PA: Beaver County Corporation for Economic Development, October.

Lendl, Iryna, Andrew R. Thomas, and Bryan Townley. 2016. "Midstream Challenges and Downstream Opportunities in the Tri-State Region." Levin College of Urban Affairs. Cleveland State University. https://engagedscholarship.csuohio.edu/cgi/viewcontent.cgi?article=2416&context=urban_facpub.

Litvak, Anya. 2015. "Shell Plans to Spend $80 Million to Clean Up Contamination at Horsehead Site." *Pittsburgh Post-Gazette*, September 22. https://www.post-gazette.com/business/powersource/2015/09/22/Paying-for-anothers-contamination-Shell-to-spend-80-million-at-Horsehead-zinc-smelter-monaca-beaver-county/stories/201509220017.

Lobao, Linda, and David Kraybill. 2009. "Poverty and Local Government: Economic Development and Community Service Provision in an Era of Decentralization." *Growth and Change* 40 (3): 418–51. https://doi.org/10.1111/j.1468-2257.2009.00489.x.

Loh, Carolyn G. 2015. "Conceptualizing and Operationalizing Planning Capacity." *State and Local Government Review* 47 (2): 134–45.

Matsco, Rebecca. 2018. "Potential Regional Economic Impacts of the Cracker Plant." Paper presented at the Symposium: Potential Regional Economic Impacts of the Ethane Cracker Plant, Washington, PA, Center for Energy Policy and Management and Local Government Academy, September 18.

Norton, Richard K. 2005. "Local Commitment to State-Mandated Planning in Coastal North Carolina." *Journal of Planning Education and Research* 24:420–36.

Pennsylvania Department of Community and Economic Development. 2015. "Keystone Opportunity Zone Program: Revitalizing the Blight and Creating Opportunities in Your Communities; Program Impact 2011–2014." https://dced.pa.gov/download/koz-report-2011-2014-f/?wpdmdl=58423.

Pennsylvania KOZ. 2017. "Keystone Opportunity Zones." Greater Hazleton Can Do. http://hazletoncando.com/new/financing/success-stories/132-Incentives/116-tax-free-koz.

Pennsylvania Legislative Budget and Finance Committee. 2009. "An Evaluation of the Keystone Opportunity Zone (KOZ) Program." Harrisburg, PA, June.

Peters, Alan, and Peter Fisher. 2004. "The Failure of Economic Development Incentives." *Journal of the American Planning Association* 70 (1): 27–37.

Romich, Eric, Nancy Bowen-Ellzey, Myra Moss, Cindy Bond, and David Civittolo. 2015. "Building Sustainability in Gas- and Oil-Producing Communities." Journal of Extension 53 (3): article 3iw1. https://archives.joe.org/joe/2015june/iw1.php.

Samuel, Leah. 2012. "Potter Township: The Forgotten Player in Bringing Shell Oil to PA." Public Source, June 25. https://www.publicsource.org/potter-township-the-forgotten-player-in-bringing-shell-oil-to-pa/.

Sokolow, Alvin D. 1981. "Community Growth and Administrative Capacity. In *Nonmetropolitan America in Transition*, edited by Amos H. Hawley and Sara Mills Mazie, chapter 19. Chapel Hill: University of North Carolina Press.

Stonesifer, Jared. 2016. "Shell Responds to Critics After Protesters Submit Petition to Potter Twp. Supervisors." December 7. https://www.timesonline.com/c86aa1b2-bcc9-11e6-a54d-3f9febf221d3.html.

Stonesifer, Jared. 2017. "A Year Later, Shell Has Forever Altered Landscape of Beaver County." June 7. http://www.timesonline.com/aaf9c622-4bb1-11e7-a3f8-dff5b6888004.html.

Stonesifer, Jared. 2018a. "Shell Pipeline Could Travel under a Dozen Local Waterways." January 25. http://www.timesonline.com/news/20180125/shell-pipeline-could-travel-under-dozen-local-waterways.

Stonesifer, Jared. 2018b. "To Shell and Back: Potter Township Supervisors Reflect on Journey of Last Six Years." May 7. http://www.timesonline.com/news/20180507/to-shell-and-back-potter-township-supervisors-reflect-on-journey-of-last-six-years.

Wereschagin, Mike, and Bill Vidonic. 2012. "Shell Picks Horsehead Plant near Monaca for Ethane Cracker." Trib Live—Tribune Total Media. March 16. https://triblive.com/x/pittsburghtrib/news/pittsburgh/s_786792.html.

Wolf, Michael Allan. 1990. "Enterprise Zones: A Decade of Diversity." In *Financing Economic Development: An Institutional Response*, edited by Richard D. Bingham, Edward W. Hill, and Sammis B. White, chapter 8, 123–41. Newbury Park, CA: Sage.

8

THE RESOURCE CONFLICT AND THE LOCAL ECONOMIC TRADE-OFFS OF FRACKING

Anna C. Osland and Carolyn G. Loh

In the quest for livable communities, planners are called to balance three aspects of sustainability: economic, environmental, and equity (Campbell 1996; Godschalk 2004). The planning ideal of balanced sustainability principles may not be achieved in practice. Using the lens of horizontal high-volume hydraulic fracturing (fracking or HVHF), we focus this chapter on the resource conflict that underscores the balance of economic-environmental issues.[1]

HVHF has potential to afford significant economic gains to communities within shale areas by providing high wage blue-collar jobs and increasing development and customers for local businesses. Scientists, economists, and planners debate the true level of benefits to local areas, since the associated local costs of fracking are also high (Christopherson and Rightor 2012; Barth 2013; Kelsey et al. 2011). HVHF could also have a potentially significant effect on the national economy by decreasing the need for oil and gas imports and driving down the price of energy; indeed, fracking could transform global energy markets as the technology to access shale gas and oil reserves transfers to other parts of the world with similar deposits (Sovacool 2014). Yet in many communities, there is a varied and engaged stakeholder group that is opposed to fracking for environmental, health, and safety reasons (Davis 2014; Schafft, Borlu, and Glenna 2013). It is genuinely difficult to weigh the economic and environmental costs and benefits of HVHF; the debate over its impacts continues even as the technology proliferates (Sovacool 2014; Howarth, Ingraffea, and Engelder 2011). It is in this context that we examine how local government officials perceive the economic costs and benefits of fracking within the larger vision of sustainability and comprehensive planning.

We use data from a survey of planners in three hundred jurisdictions in four states (Pennsylvania, North Dakota, Louisiana, and Colorado) that are partially or totally located within shale basins. We ask (1) if perceived economic benefit is equal between residents and local businesses, (2) if there are differences and how we can explain them, and (3) if communities that have benefited economically from fracking have adopted growth management tools to address sustainability. We analyze the data using descriptive statistics and nonparametric tests.

Based on the survey results, we find that respondents view fracking as having a smaller benefit for citizens than for businesses. This seems to be because respondents perceive fracking to have a negative impact on housing affordability, but a positive impact for overall economic development. We find that the relationship between the perception of economic benefit from fracking and adoption of growth management tools to address fracking externalities is mixed. There was no statistical relationship between the perception of economic benefit from fracking and the majority of tools local governments used to address fracking, yet there was a statistically significant relationship with adoption of tools that address fracking by-products. We expect that this relationship demonstrates that while the economic benefit from fracking does lead local governments to accept fracking for the potential revenue, the lack of economic benefit from fracking by-products leads them to restrict the use and location of these activities.

Fracking, Planning, and the Resource Conflict

Planners have long been called to create a balance between policies that encourage economic development and those that protect the environment, with development often winning out in what Campbell calls the "resource conflict" (Godschalk 2004; Campbell 1996). As Campbell explains, "Our historic tendency has been to promote the development of cities at the cost of natural destruction: to build cities we have cleared forests, fouled rivers and the air, leveled mountains" (Campbell 1996, 296). Planners have engaged in considerable soul-searching as to the nature of the resource conflict and what it means for sustainability. Connelly (2007) argues that even defining the sustainability at the center of Campbell's planner's triangle is an inherently political act that requires the definer to make normative choices. For example, focusing only on the resource conflict, what level of natural resource consumption versus what level of economic growth would be considered sustainable? What about those who would advocate a "no-growth" strategy? This idea of sustainability is "unreachable in any complete and final way yet ever present as a guiding pole in relation to which planners can orient themselves" (Connelly 2007, 263). For example, sustainability need not prioritize *any* objectives that benefit humanity, yet Campbell's definition places social justice and economic development on equal footing with environmental protection (Connelly 2007, 267).[2]

From a more cynical viewpoint, emphasis on sustainability can be seen as a cover for practices that companies and governments would engage in anyway to operate more efficiently or cost-effectively (Luke 2005). For example, reducing the amount of resources a company consumes to manufacture its products helps the environment but also reduces costs. If it were more expensive to reduce the amount of resources consumed, the company would be less likely to take this sustainability measure. In the public sector, plans that state a commitment to sustainability are no more likely to include specific implementation measures that would actually lead to more sustainable outcomes than those that do not mention it, illustrating the difficulty planners have in translating any definition of sustainability into concrete action (Berke and Conroy 2000). Rather than attempting to resolve the resource conflict or directly confronting the trade-offs it would entail, policy makers may task different agencies with the two conflicting goals (Thacher and Rein 2004); for example, having planners engage in economic development and leaving environmental protection to another agency at the local or state level. The contemporary debate around high-volume hydraulic fracturing exemplifies this resource conflict, at both the national and local levels.

Proponents of high-volume hydraulic fracturing, in which a mixture of water, sand, and chemicals is pumped thousands of meters underground to extract oil and natural gas in shale rock formations, point out that hydraulic fracturing, in combination with other alternative sources of energy, is setting the United States up to be a net energy exporter (Meyer 2012; Negro 2012; Nolon 2012; Boudet et al. 2014; Blackwill and O'Sullivan 2014). Fracking also makes the United States less dependent on coal, with its impacts on local air pollution and greenhouse gas emissions (Inglesby et al. 2012). Thus, supporters of fracking argue that it is both more economically and more environmentally sustainable. Opponents of fracking raise concerns about large-scale impacts on water quality, particularly the potential for groundwater and drinking water well contamination and the permanent removal of contaminated water from the water cycle (Vidic et al. 2013; Rabe and Hampton 2015). Some also argue that greenhouse gas reductions associated with increased natural gas use are overstated since the fracking process releases methane into the air (Peduzzi and Harding Rohr Reis 2013; Howarth

2014). As a separate public safety issue, the evidence is also becoming increasingly clear that fracking can contribute to seismic activity (Holland 2011).

Despite its proponents' claims, the emerging scientific consensus seems to be that fracking is not an environmentally neutral practice, although the full extent of its environmental impacts is not yet understood. This lack of long-term information makes it difficult to weigh costs and benefits, and differing assumptions about accident rates and future oil prices lead to vastly different conclusions about whether fracking should be stopped or expanded. This divide is well illustrated in a *Nature* point-counterpoint article from several years ago with scientists passionately arguing on either side of the debate (Howarth, Ingraffea, and Engelder 2011). In his review of the fracking literature, Sovacool (2014, 249) says that because of the difficulty and complexity involved in measuring and weighing costs and benefits, fracking "presents policymakers, planners, and investors with a series of pernicious trade-offs and tough choices." Recurring accidents, including explosions, leaks, and spills, add more weight to the environmental harm side of the debate, although others point out that such accidents are caused by human error, which is present and theoretically reducible in any extraction technology (Howarth, Ingraffea, and Engelder 2011).

At the local level, in many places there is a great deal of support for fracking, for good reasons (Davis 2012). The gas and oil industry has made claims that fracking will lead to the addition of millions of jobs to the US economy (Efstathiou 2012). In some areas of the country with a long history of gas and oil extraction, fracking does not seem qualitatively different than other forms of extraction that have been performed for years (Howarth, Ingraffea, and Engelder 2011). In these areas, the petroleum industry is a large part of the economy, and new extraction techniques can mean more and better jobs (Holdman 2013; Reindl 2014). In Texas, oil and gas revenues are a significant source of school funding (Davis 2012). Even in parts of the country that have a more conflicted relationship with gas and oil development, the significant economic benefits fracking brings may mitigate environmental concerns. For example, Pennsylvania has a long history of environmental damage from oil and gas exploitation (Wilber 2015), but Act 13 impact fees from fracking can fund a host of local government projects (Sacavage 2013); we found in chapter 5 that many municipalities use them for common areas. Yet many researchers think the jobs claims are significantly overstated. Analysis of job growth in Texas, Pennsylvania, and Ohio shows that most states add jobs in the low tens of thousands rather than the hundreds of thousands reported by industry projections (Mauro et al. 2013; Ritholtz 2015; Weinstein and Partridge 2011). In addition, the jobs fracking creates fluctuate significantly along with oil prices: the US Bureau of Labor Statistics (2017) estimates that the oil and gas extraction industry lost over twenty thousand jobs between January 2015 and January 2017. Industries that support oil and gas activities lost many thousands more. Statewide trusts funded by oil and gas severance (extraction) taxes can help provide a steady stream of income and mitigate the boom-bust cycle (Saha and Muro 2016). Some portion of severance taxes may be set aside specifically to mitigate negative environmental externalities (Rabe and Hampton 2015).

The literature indicates that there is considerable difficulty in finding a policy balance between environmental protection and economic development. Debates around fracking clearly illustrate this challenge. Even if planners and other policy makers were completely clear on where they found a balance within the resource conflict, it would be difficult to assess whether fracking met their criteria for sustainability. The environmental costs of fracking are still under investigation and its long-term impacts are not known. The economic benefits of fracking are contested and look very different at different geographic scales and under different assumptions about the future price of oil. Yet local officials are still compelled to make policy decisions about fracking, including the decision about whether to have any policy at all, given many states' laws preempting local governments from making policy around fracking (Loh and Osland 2016; Cook 2014). They must have an opinion about how beneficial fracking will be to the residents and business owners in their communities, how much harm it might do, and how to weigh the trade-offs involved. Thus, we ask our three research questions. First, we ask if perceived economic benefit is equal between citizens and local businesses: if local officials think fracking benefits their communities, whom do they think it benefits? Second, if there are differences between perceived benefits for citizens and businesses, what factors influence the perception of those differences? Finally, we ask if communities that have benefited economically from fracking have adopted growth management tools to address sustainability. Here we are looking for evidence of local officials attempting to find a balance within the resource conflict. In the next sections, we describe our survey and analytical methodology and results, then discuss the implications of those results.

Methodology

Study Area

We investigated our three research questions using a combined email/mail/ phone survey of municipalities in four states. To draw our survey sample, we used a 2011 GIS layer of shale (both oil and gas) plays from the Energy Information Administration. Within those areas, we surveyed all counties (parishes in Louisiana) in the four study states. We also surveyed all places (and county subdivisions in Pennsylvania) with more than 10,000 people as of the 2010 US Census. In Colorado and North Dakota, we also surveyed all places with more than 5,000 people, as those states are much less densely populated, especially in the areas containing the shale plays.[3] This resulted in a sample of about three hundred communities, of which 49 percent responded to the survey. For detailed information about our survey methodology, please see our earlier article on local land use responses to hydraulic fracturing (Loh and Osland 2016).

Data Analysis

In our data analysis we were interested in investigating three research questions: (1) if perceived economic benefit is equal between citizens and local businesses, (2) what factors are associated with perception of economic benefit for both citizens and businesses, and (3) if communities that have benefited economically from fracking have adopted growth management tools to address sustainability. To help answer these questions we asked survey respondents a series of questions about the impact of fracking on various aspects of community life. For each question, the respondent could say that fracking had made that aspect of community life better (3), worse (1), or that there had been no change (2). To evaluate the equality of the fracking benefit for citizens and businesses, we asked planners to evaluate the effect (better, no change, worse) on citizens and businesses, respectively. Since our data is ordinal and not normally distributed, we test the equality of benefit between these two groups using a nonparametric Wilcoxon rank-sum test.

We then evaluated the factors that influence perception of economic benefit for citizens and businesses. For this analysis we assessed the relationship between housing and development issues and their impact on citizens and businesses. We used Spearman's rank test to evaluate the correlation between each of these variables.

We used the same survey data to analyze our third research question, if communities that benefited economically from fracking have adopted growth management tools to address sustainability. In the survey, we asked respondents which of twenty-two land use tools related to fracking their communities had adopted (Loh and Osland 2016). For the purposes of this chapter, we focus on a subset of those tools that are directly related to minimizing environmental impacts of fracking (as opposed to protecting humans from potential impacts of

TABLE 8.1 Land use tools aimed at fostering environmental sustainability

TOOL	PERCENT OF GOVTS. ADOPTED
Special hazard ordinance for fracking activity	9
Restrictions on locations of industrial land uses	30
Special use permit requirements for drilling sites	26
Zoning for drilling by-product disposal	16
Restrictions on fracking activity near water supply reservoirs	11
Watershed protection ordinance	10
Moratoria on fracking activity*	3
Restrictions on disposal of drilling by-products	15
Impact fees for fracking	14
Environmental impact statement for fracking sites	8
Stringent conditions on drilling permits	8

* At the time of the survey, several Colorado communities were attempting to enact moratoria on fracking and were unable to answer the survey for legal reasons.

Source: Authors' calculations from survey.

fracking).[4] Table 8.1 shows the land use tools we considered to be aimed at foster-ing environmental sustainability. Each of the land use tool variables is a dummy that records either adopted or not adopted. For informational purposes, we also include the percentage of responding communities that reported adopting each tool. The percentage of communities using any of the land use tools to address the impacts of fracking is quite low.

To investigate the economic implications of fracking on a community, we constructed two index variables for housing affordability and for economic development activities. The "housing affordability" variable was constructed from two questions about housing affordability and rental unit availability (see table 8.2 for more details). This index variable was created to capture the impact of fracking on household budgets. The Cronbach's alpha statistic for this index was 0.88. The "economic development activity" index variable represents the amount of economic activity generated by fracking, which includes effect on tax base, demand for new commercial development, and demand for new residential development (table 8.2). The Cronbach's alpha statistic for this index variable was 0.73. Each index variable was constructed by adding together the values of the component variables and dividing by two for the housing index and by three for the economic development activity index. We use the nonparametric Spearman's rank test to evaluate the correlation of adoption of each of the eleven tools in table 8.1, with economic indicators for impact on businesses, citizens, housing, development, and years a community has had fracking.

TABLE 8.2   Survey questions used to create index variables

Housing affordability index

Has fracking had an effect on rental unit availability in your jurisdiction? (Better / Worse / No change)

Has fracking had an effect on housing affordability in your jurisdiction? (Better / Worse / No change)

Economic development index

Has fracking had an effect on the tax base in your jurisdiction? (Better / Worse / No change)

Has fracking had an effect on demand for new commercial development in your jurisdiction? (Better / Worse / No change)

Has fracking had an effect on demand for new residential development in your jurisdiction? (Better / Worse / No change)

Findings

We find that few respondent communities felt fracking had a negative economic impact on either local businesses or citizens (table 8.3). Using the Wilcoxon rank-sum test, we do find statistically significant differences between the two groups (z=3.45, p<0.01), indicating that planners and other local officials see more economic benefit for businesses than for citizens.

While few planners observed that fracking had a negative economic effect on either local businesses or citizens, neither did they uniformly perceive that fracking provided economic gains for all. Table 8.4 illustrates the differences in perception of economic effect by state. Using the Wilcoxon rank-sum test, we did not find statistically significant differences in perception of economic gain in any state, except Pennsylvania (z=2.31, p<0.05). In Pennsylvania, planners reported a higher benefit from fracking for businesses than for citizens.

We find that housing issues are statistically significant and negatively correlated with a perception of impact of fracking on businesses and on citizens (table 8.5). This correlation suggests that a perception of fracking's positive economic effect on both businesses and citizens is associated with a combination of reduced housing affordability and rental unit availability. In contrast, we find that economic development activity is statistically significant and positively correlated with a perception of fracking's impact on businesses and citizens, suggesting that a positive perception of fracking's impact in a community is correlated with additional community-level economic activity.

Although we find a statistically significant difference between the economic effect of fracking on businesses and citizens, we did not find enough variation in our survey responses to test for differences in adoption of growth management tools among communities with lower economic benefit to citizens than

TABLE 8.3 Economic effect of fracking on businesses and citizens

	ECONOMIC EFFECT		
	BETTER	NO CHANGE	WORSE
Businesses (n=107)	54%	46%	0%
Citizens (n=106)	40%	58%	2%

TABLE 8.4 Economic effect of fracking on businesses and citizens, by state

Citizens (n=106)

ECONOMIC EFFECT	COLORADO	LOUISIANA	NORTH DAKOTA	PENNSYLVANIA
Worse	0%	0%	11%	0%
No change	70%	57%	28%	62%
Better	30%	43%	61%	38%
Total	100%	100%	100%	100%

Businesses (n=107)

ECONOMIC EFFECT	COLORADO	LOUISIANA	NORTH DAKOTA	PENNSYLVANIA
Worse	0%	0%	0%	0%
No change	60%	25%	28%	49%
Better	40%	75%	72%	51%
Total	100%	100%	100%	100%

TABLE 8.5 Correlation between perception of fracking's impact on businesses and citizens with housing affordability and economic development activity (n=106)

	HOUSING AFFORDABILITY INDEX	ECONOMIC DEVELOPMENT ACTIVITY INDEX
Impact on business	−0.47***	0.64***
Impact on citizens	−0.43***	0.61***

*p<0.1, **p<0.05 ***p<0.01

businesses. In fact, very few communities reported that fracking had a negative economic impact on either their citizens or businesses.

As shown in table 8.6, we find few economic indicators were statistically significant and strongly correlated with adoption of sustainability tools to

TABLE 8.6   Correlation of economic indicators on adoption of sustainability tools that address fracking

LAND USE TOOL	ECONOMIC INFLUENCE				
	IMPACT ON BUSINESSES	IMPACT ON CITIZENS	HOUSING AFFORDABILITY INDEX	ECONOMIC DEVELOPMENT ACTIVITY INDEX	YEARS OF FRACKING
Special hazard ordinance for fracking activity	—	—	—	—	—
Restrictions on locations of industrial land uses	—	—	—	—	–0.21**
Special use permit requirements for drilling sites	—	—	—	—	—
Zoning for drilling by-product disposal	—	0.22**	—	0.22**	0.18*
Restrictions on fracking activity near water supply reservoirs	—	—	—	—	—
Watershed protection ordinance	—	—	—	—	—
Moratoria on fracking activity	—	—	—	—	—
Restrictions on disposal of drilling by-products	—	—	–0.18*	0.18*	—
Impact fees for fracking	—	—	—	0.22**	—
Environmental impact statement for fracking sites	—	—	0.18*	—	—
Stringent conditions on drilling permits	—	—	—	—	—

*p<0.1, **p<0.05 ***p<0.01

address fracking. Interestingly, restrictions on disposal of drilling by-products were negatively correlated with the housing affordability index, yet positively correlated with the economic development activity index. This finding would seem to indicate that as positive economic benefits from fracking increase, communities show less interest in receiving fracking by-products that do not convey the economic benefits associated with fracking. Adoption of zoning for drilling by-product disposal was statistically significant and positively correlated with both the economic development activity index and the effect on citizens. Perhaps unsurprisingly, we found a statistically significant and positive correlation between the economic development activity index and adoption of impact fees.

The Resource Conflict Revisited

We initially expected to see planners and officials in communities with fracking to be concerned with the trade-offs involved in the resource conflict between environmental protection and economic development. Our data did not indicate that local officials are worried about this conflict; in fact, there were so few communities where respondents perceived any negative impacts of fracking that we were unable to test differences in trade-offs between impacts on citizens and impacts on businesses (we had expected to see a difference in economy/environment trade-offs between communities where there was a higher benefit to citizens than businesses, but that is just not the case). The environmental risks for residents in fracking areas include water contamination, air pollution, chemical spills, and fires, but either local officials have not experienced these problems, or they give these potential negative effects less weight than they do the tangible positive economic impacts. It is also possible that our data set is skewed toward positive attitudes toward fracking's impact on citizens and businesses since we are aware that several communities involved in lawsuits could not answer the survey, and there may be additional communities in similar situations of which we are not aware. Regardless, our data suggest that most of our respondent communities are coming down squarely on the economic growth side of the resource conflict, whether they have thought through that decision or not.

The academic literature on fracking we explored at the beginning of this chapter reflects a great deal of ambiguity about the costs and benefits of fracking. Some scientists studying this technology even go so far as to call for moratoria on fracking until we understand more about the long-term environmental consequences. The positive perceptions of fracking's economic impact we found among local officials, coupled with the lack of attention to growth management or other environmental controls, suggests that out in the "real world" there is considerably less concern for the environmental corner of the planner's triangle. Of course, one way to resolve the resource conflict is to fully prioritize the economy.

Communities frequently discount the potential consequences from hazards when making planning decisions that reduce vulnerability (Burby 2006; Osland 2013), and the limited interest we found in addressing the resource conflict may illustrate this behavior. In previous analysis of our survey data, we found that local planning capacity plays a significant role in determining whether a community attempts to regulate fracking at the local level, although most communities do not regulate it at all, and the ones that do regulate it do not do so very strongly (Loh and Osland 2016). We know from other research that planning capacity helps increase plan quality and implementation as well (Loh 2015). Moreover, planners can build community capacity through collaborative partnerships among technical experts in order to help transfer knowledge from specialists to community members (Osland 2015). It seems that there is a strong role for planners here, both in facilitating collaboration and in educating local officials about the trade-offs they are implicitly making when they make policy decisions around fracking.

The negative correlation between the restrictions on locations of industrial land uses and the total years a community has had fracking may suggest that communities where fracking is a more recent phenomenon may be attempting to affect the location of fracking activity. Locations with more fracking experience may be more interested in increasing the opportunities to continue the economic gains from fracking by reducing the barriers to industrial development. These communities may also have a long history of oil and gas extraction, and residents and officials there may be more comfortable with the environmental and economic trade-offs of the industry in general. While the public continues to develop its opinion about fracking (Boudet et al. 2014), there may be more tolerance for the risks associated with fracking in communities that have experience with the oil and gas industry than in those without this type of involvement.

We do find some differences between communities' responses to fracking and their responses to fracking by-products. It is possible that communities that have more fracking also have more by-products, and those communities are more interested in regulation. Since the public displays a lower perception of risk from commonly encountered and voluntary activities (Slovic 1987), it may be that familiarity with oil and gas extraction leads to acceptance of the risks from the extraction aspect of fracking, but does not extend to support for by-product disposal activities with less well known consequences. It could also be the case that as communities see an increased impact from fracking, they are less tolerant of accommodating by-products. A last possibility is that communities are willing to have fracking, which comes with economic benefits, but less willing to have disposal, which carries environmental risk but does not contribute the same economic benefits. The regulatory environment for regulating fracking by-products may be different in some states than that for regulating fracking itself, empowering local governments to reject disposal sites even while they must accept extraction wells (Zwick 2018). This may be an indirect way for local governments to influence fracking activity, as without disposal wells, extraction cannot take place.

Although we did not see any testable difference in policy responses based on differential benefits to citizens and businesses, we do see a difference in perception of benefit between citizens and businesses, with businesses benefiting more. Additionally, the perception of economic gain was not uniform between states. Planners in Colorado and Pennsylvania recorded more limited perception of benefit for citizens than planners in Louisiana and North Dakota. Only in Pennsylvania was there a statistically significant difference in the perception of economic benefit between citizens and businesses. Planners perceived that fracking positively affected businesses, while citizens were neither positively nor negatively affected. The data make it clear that many communities with HVHF are experiencing boomtown conditions. Although most respondents reported that the economic impact of fracking on communities is positive, that positive economic impact was significantly correlated with housing conditions that make life more expensive for residents. However, these negative impacts on citizens are clearly not strong enough for most respondents to consider fracking an overall negative for the community.

While the short-term local economic benefits of HVHF are relatively clear—an increase in high-paying jobs, ability to charge higher rents and sell more houses, increased patronage for local restaurants and other businesses—as a long-term economic development strategy, reliance on HVHF is unlikely to be successful for any one community. Production from individual wells drops off after a few years, and the industry moves to new areas. Oil prices are volatile, and fossil fuels face long-term competition from and government incentives for renewable energy sources. Given the volatile and likely short-term nature of the community economic benefit, this chapter highlights the importance of collaboration for achieving economic and environmental planning goals together as part of a strategy of reaching community visions of sustainability.

Consideration for how collaboration fits within visions of sustainability is especially relevant given the regional scale potentially affected by fracking accidents. Fracking may be part of a long-term economic sustainability strategy, but it should not be the only or maybe not even the main piece, and the risk of long-term environmental damage should be factored in. Collaborative practices might help planners find a middle ground.

Next Steps

Fracking is a highly controversial activity, as exemplified by the mixed conclusions in the academic literature on costs and benefits. Having planners legally unable to respond to questions about fracking suggests that this is an ongoing controversy within a rapidly changing policy environment. Although we suspect fracking, like other boom-bust resource extraction, is not a sustainable economic development strategy, a future survey could explicitly explore the potential for long-term economic gain from fracking by asking communities with longer histories of fracking about their experiences. Such a research approach would help illuminate how local planners and officials define sustainability, and whether and in what ways they attempt to steer their communities toward sustainable land use and economic choices. We also believe that the case study approach from the previous three chapters in Part II of this volume can help to illuminate more specifically some of the questions that the survey data could not answer, such as how and how much local officials and residents consider the resource conflict when making policy decisions, which stakeholders are involved in setting policy around fracking, and more specific information about who in the community is benefiting from fracking.

Notes

1. See chapter 5 for more on the resource conflict.

2. Note that in this chapter we focus almost entirely on the resource conflict, yet the question of equity is always present, as the costs and benefits of fracking, like those of other technologies, are rarely equitably distributed.

3. At the time of the survey, several Colorado communities were attempting to enact moratoria on fracking and were unable to answer the survey for legal reasons.

4. There are a few tools that are clearly aimed at protecting both humans and the natural environment; we include them here.

References

Barth, Jannette M. 2013. "The Economic Impact of Shale Gas Development on State and Local Economies: Benefits, Costs, and Uncertainties." *New Solutions: A Journal of Environmental and Occupational Health Policy* 23 (1): 85–101.

Berke, Philip R., and Maria Manta Conroy. 2000. "Are We Planning for Sustainable Development? An Evaluation of 30 Comprehensive Plans." *Journal of the American Planning Association* 66 (1): 21–33. https://doi.org/10.1080/019443600089 76081.

Blackwill, Robert D., and Meghan L. O'Sullivan. 2014. "America's Energy Edge." *Foreign Affairs* 93 (2): 21–22.

Boudet, Hilary, Christopher Clarke, Dylan Bugden, Edward Maibach, Connie Roser-Renouf, and Anthony Leiserowitz. 2014. "'Fracking' Controversy and Communication: Using National Survey Data to Understand Public Perceptions of Hydraulic Fracturing." *Energy Policy* 65:57–67.

Burby, R. J. 2006. "Hurricane Katrina and the Paradoxes of Government Disaster Policy: Bringing About Wise Governmental Decisions for Hazardous Areas." *Annals of the American Academy of Political and Social Science* 604 (1): 171–91.

Campbell, Scott. 1996. "Green Cities, Growing Cities, Just Cities? Urban Planning and the Contradictions of Sustainable Development." *Journal of the American Planning Association* 62 (3): 296–312.

Christopherson, Susan, and Ned Rightor. 2012. "How Shale Gas Extraction Affects Drilling Localities: Lessons for Regional and City Policy Makers." *Journal of Town and City Management* 2 (4): 350–68.

Connelly, Steve. 2007. "Mapping Sustainable Development as a Contested Concept." *Local Environment* 12 (3): 259–78. https://doi.org/10.1080/1354983060118 3289.

Cook, Jeffrey J. 2014. "Who's Regulating Who? Analyzing Fracking Policy in Colorado, Wyoming, and Louisiana." *Environmental Practice* 16 (2): 102–12.

Davis, Charles. 2012. "The Politics of 'Fracking': Regulating Natural Gas Drilling Practices in Colorado and Texas." *Review of Policy Research* 29 (2): 177–91.

Davis, Charles. 2014. "Substate Federalism and Fracking Policies: Does State Regulatory Authority Trump Local Land Use Autonomy?" *Environmental Science and Technology* 48 (15): 8397–8403. https://doi.org/10.1021/es405095y.

Efstathiou, Jim. 2012. "Fracking Will Support 1.7 Million Jobs, Study Shows." Bloomberg, October 23. http://www.bloomberg.com/news/articles/2012-10-23/fracking-will-support-1-7-million-jobs-study-shows.

Godschalk, David R. 2004. "Land Use Planning Challenges: Coping with Conflicts in Visions of Sustainable Development and Livable Communities." *Journal of the American Planning Association* 70 (1): 5–13.

Holdman, Jessica. 2013. "Oil-Related Companies on North Dakota Hiring Blitzes." Casper Star Tribune, December 29. http://trib.com/business/energy/oil-related-companies-on-north-dakota-hiring-blitzes/article_f2074093-65e4-5b01-895c-8896c2b39f8e.html.

Holland, Austin. 2011. "Examination of Possibly Induced Seismicity from Hydraulic Fracturing in the Eola Field, Garvin County, Oklahoma." Oklahoma Geological Survey Open-File Report OF1-2011. http://www.ogs.ou.edu/pubsscanned/openfile/OF1_2011.pdf.

Howarth, Robert W. 2014. "A Bridge to Nowhere: Methane Emissions and the Greenhouse Gas Footprint of Natural Gas." *Energy Science & Engineering*. https://doi.org/10.1002/ese3.35.

Howarth, Robert W., Anthony Ingraffea, and Terry Engelder. 2011. "Natural Gas: Should Fracking Stop?" *Nature* 477 (7364): 271–75.

Inglesby, Tommy, Rob Jenks, Scott Nyquist, and Dickon Pinner. 2012. "Shale Gas and Tight Oil: Framing the Opportunities and Risks." McKinsey on Sustainability and Resource Productivity. https://www.mckinsey.com/~/media/mckinsey/dotcom/client_service/Sustainability/PDFs/McK%20on%20SRP/SRP_04_Shale%20gas.ashx.

Kelsey, T., M. Shields, J. R. Ladlee, and M. Ward. 2011. "Economic Impacts of Marcellus Shale in Pennsylvania: Employment and Income in 2009." Marcellus Shale Education and Training Center. Penn State Extension and Pennsylvania College of Technology. https://www.pct.edu/files/imported/business/shaletec/docs/EconomicImpactFINALAugust28.pdf.

Loh, Carolyn G. 2015. "Conceptualizing and Operationalizing Planning Capacity." *State and Local Government Review* 47 (2): 134–45. https://doi.org/10.1177/016 0323x15590689.

Loh, Carolyn G., and Anna C. Osland. 2016. "Local Land Use Planning Responses to Hydraulic Fracturing." *Journal of the American Planning Association* 82 (3): 222–35. https://doi.org/10.1080/01944363.2016.1176535.

Luke, Timothy W. 2005. "Neither Sustainable nor Development: Reconsidering Sustainability in Development." *Sustainable Development* 13 (4): 228–38. https://doi.org/10.1002/sd.284.

Mauro, Frank, Michael Wood, Michele Mattingly, Mark Price, Stephen Herzenberg, and Sharon Ward. 2013. "Exaggerating the Employment Impacts of Shale Drilling: How and Why." Multi-State Shale Research Collaborative. http://www.fiscalpolicy.org/wp-content/uploads/2013/11/MSSRC-Employment-Impact-11-21-2013.pdf.

Meyer, Andrew. 2012. "'Get the Frack Out of Town': Preemption Challenges to Local Fracking Bans in New York." *Columbia Journal of Environmental Law*. Field Reports 22 April 2012. https://perma.cc/UVR8-DQCN.

Negro, S. 2012. "Fracking Wars: Federal, State and Local Conflicts over the Regulation of Natural Gas Activities." *Zoning and Planning Law Report* 35 (2): 1–15.

Nolon, John R. 2012. "Hydrofracking: Disturbances Both Geological and Political: Who Decides?" Urban Lawyer 44 (3). https://digitalcommons.pace.edu/lawfaculty/834.

Osland, Anna C. 2013. "Using Land-Use Planning Tools to Mitigate Hazards: Hazardous Liquid and Natural Gas Transmission Pipelines." *Journal of Planning Education and Research* 33 (2): 141–59.

Osland, Anna C. 2015. "Building Hazard Resilience through Collaboration: The Role of Technical Partnerships in Areas with Hazardous Liquid and Natural Gas Transmission Pipelines." *Environment and Planning A* 47 (5): 1063–80.

Peduzzi, Pascal, and Ruth Harding Rohr Reis. 2013. "Gas Fracking: Can We Safely Squeeze the Rocks?" *Environmental Development* 6:86–99.

Rabe, Barry G., and Rachel L. Hampton. 2015. "Taxing Fracking: The Politics of State Severance Taxes in the Shale Era." *Review of Policy Research* 32 (4): 389–412. https://doi.org/10.1111/ropr.12127.

Reindl, J. C. 2014. "Fracking Boom Has Been a Jobs Boon for North Dakota." *Detroit Free Press*, October 12. http://www.freep.com/story/money/business/michigan/2014/10/12/north-dakota-energy-boom-michigan-jobs/17044837/.

Ritholtz, Barry. 2015. "Energy, Texas, and Job Creation." *Bloomberg*, January 26. http://www.bloomberg.com/view/articles/2015-01-26/energy-texas-and-flawed-analysis-of-job-growth.

Sacavage, Krystle J. 2013. "Overview of Impact Fee Act: Act 13 of 2012." Pennsylvania Public Utility Commission.

Saha, Devashree, and Mark Muro. 2016. "Permanent Trust Funds: Funding Economic Change with Fracking Revenues." Advanced Industries Series. Washington, DC: Brookings Institution. https://www.brookings.edu/research/permanent-trust-funds-funding-economic-change-with-fracking-revenues.

Schafft, Kai A., Yetkin Borlu, and Leland Glenna. 2013. "The Relationship between Marcellus Shale Gas Development in Pennsylvania and Local Perceptions of Risk and Opportunity." *Rural Sociology* 78 (2): 143–66.

Slovic, Paul. 1987. "Perception of Risk." *Science* 236 (4799): 280–85. https://doi.org/10.1126/science.3563507.

Sovacool, Benjamin K. 2014. "Cornucopia or Curse? Reviewing the Costs and Benefits of Shale Gas Hydraulic Fracturing (Fracking)." *Renewable and Sustainable Energy Reviews* 37:249–64.

Thacher, David, and Martin Rein. 2004. "Managing Value Conflict in Public Policy." *Governance* 17 (4): 457–86. https://doi.org/10.1111/j.0952-1895.2004.00254.x.

US Bureau of Labor Statistics. 2017. "Employment, Hours, and Earnings from the Current Employment Statistics Survey (National)." In All Employees (in

Thousands), Oil and Gas Extraction (Seasonally Adjusted). BLS.gov. https://www.bls.gov/iag/tgs/iag211.htm.

Vidic, R. D., S. L. Brantley, J. M. Vandenbossche, D. Yoxtheimer, and J. D. Abad. 2013. "Impact of Shale Gas Development on Regional Water Quality." *Science* 340 (6134): 1235009.

Weinstein, Amanda L., and Mark D. Partridge. 2011. "The Economic Value of Shale Natural Gas in Ohio." Swank Program in Rural-Urban Policy Summary and Report. Ohio State University Extension. https://aede.osu.edu/sites/aede/files/publication_files/Economic%20Value%20of%20Shale%20FINAL%20Dec%202011.pdf.

Wilber, Tom. 2015. *Under the Surface: Fracking, Fortunes, and the Fate of the Marcellus Shale*. Updated ed. Ithaca, NY: Cornell University Press.

Zwick, Austin. 2018. "The Public Finance Challenges of Fracking for Local Governments in the United States." IMFG Papers on Municipal Finance and Governance 38: 1–26. http://hdl.handle.net/1807/87370.

Part III
ECONOMIC IMPACT

LOCAL LABOR MARKETS AND SHALE GAS

Frederick Tannery and Larry McCarthy

The ability to extract commercial quantities of natural gas by horizontal drilling combined with hydraulic fracturing has opened large new reserves and substantially increased production. Shale gas now accounts for 60 percent of dry gas in the United States and is responsible for the United States becoming the largest energy producer in the world. US shale gas production has increased despite falling energy prices as output increased from 7,994 billion cubic feet in 2011 to 18,589 billion cubic feet in 2018 as shale has become a low-cost source of energy compared to deep-water drilling and other tertiary extraction methods. These developments have had a dramatic impact on the entire energy sector with substantial spillover effects on regions in which the gas is located, and on the entire consumer community as the market power of producers has been diminished with sharply falling prices as a consequence.

One notable region that is producing substantial quantities of natural gas is in the Marcellus Shale in Pennsylvania. Pennsylvania has the largest shale gas production and the large majority comes from rural counties. Production in Pennsylvania played a lead role in the national surge as production rose from 1,068 billion cubic feet in 2011 to 5,365 billion cubic feet in 2017. Five of the nine largest-producing counties are in sparsely populated northern regions that border on New York; and three other counties, also in rural areas, share borders with these five.[1]

As gas production represents a substantial boom to these relatively small rural economies, this chapter assesses the impact of gas production in regions in which gas is concentrated. Exploiting the differences in gas production to estimate how the extraction of gas affects local employment and earnings, we compare the outcomes in counties that have large quantities of gas with others that have little or no gas. Estimates of the effect of gas production on local employment and wages have important effects on local policy makers, who must decide on how to incentivize the production along with taxing production to compensate for the negative externalities of possible environmental damage and highway damage caused by heavy trucks transporting wastewater. Measuring the effect of shale gas production requires constructing a suitable comparison group that is similar to counties with gas production. Since this study only considers Pennsylvania counties, there is a strong possibility there could be spillover effects of gas production on low or no gas counties, which produces a downward bias in our estimates.

We measure gas production in several ways. First, we compare high gas producing counties to others that have little or no gas; we also use the semiannual change in gas production and estimate the impact of changes in gas production on employment and wages. Finally, since gas production requires exploration and drilling activity before gas is extracted, we use the entire gas production in the county over the study period. Our empirical approach follows Weber (2012) who estimated the impact of the natural gas boom on counties in Colorado, Texas, and Wyoming from 1999 through 2008, before gas prices collapsed. Our approach also follows Black, McKinnish, and Sanders (2005), who estimate the effects of the boom and bust in the coal industry on local labor markets in largely rural counties in Kentucky, Ohio, Pennsylvania, and West Virginia. Our contribution is a complementary study of the impact of a resource boom that includes a region closer to a larger population center and covers a particularly sharp recession over a period when gas prices peaked and began to decline. Furthermore, location differences between treatment and control counties have idiosyncratic contrasts that would make comparison vary from the other studies.

Literature, Data, and Methodology

The results of research on the impact of resource booms vary depending upon the location and size of the boom. The theory developed by Corden and Neary (1982) finds that in a small open economy, resource booms lead to higher local wages and exchange rates that adversely affect domestic industries (i.e., Dutch Disease). Research regarding booms within different countries uses time-series data to measure the effect of a resource boom, compared to a sometimes difficult-to-construct comparison group. Papyrakis and Gerlagh (2007) and James and Aadland (2011) find that resource-dependent economies grew more slowly, while Michaels (2010) shows that counties located on major oil fields between 1890 and 1990 had a high per capita income. Marchand (2012) estimates that the energy boom in western Canada produced higher employment and wages between 1996 and 2006; and Weber (2012), in summarizing the resource boom literature, notes that the adverse impact of an energy boom on economic growth is reversed when the boom sector is a small part of a larger economy. Studies with a long-run perspective consider both the boom and bust cycle: both Black, McKinnish, and Sanders (2005) and Jacobsen and Parker (2016) find that the growth in local economies during the boom is more than offset by the fall during the bust. Weber (2012) also finds a modest boost to employment and wages over the boom and bust cycle, but his estimates are much smaller than the estimates from input-output models, for example, Considine, Watson, and Blumsack (2010).

This research uses the Quarterly Census of Employment and Wages (QCEW) data set that is maintained by the Bureau of Labor Statistics (BLS). The QCEW is a remarkably rich data set and is well suited for this task. The data comes from employment and earnings data collected by the Pennsylvania Department of Labor and Industry (L&I) from the QCEW data collection effort of the BLS. Employers with workers covered by the unemployment insurance program (UI) are required to report the earnings paid to workers, which is used to determine the eligibility for and the amount of UI benefits and the UI tax liability of firms. Over 98 percent[2] of all workers are covered by the UI program. Earnings and employment are reported at each separate establishment, aggregated over industry and county to report employment in each month of the quarter, average earnings, and the number of establishments in each quarter. Since employer data is subject to disclosure rules, no information is provided if there are fewer than three employers in the county in a particular industry or if a single employer contains 80 percent or more of the industry's employment.[3] Our study period begins with 2001 and ends with the second quarter of 2015. The period covers two recessions including the sharpest business contraction since the 1930s. Gas production data comes from the Pennsylvania Department of Environmental Protection, which reports gas from unconventional wells (horizontal drilling). The data is reported annually before 2010 and then on a semiannual basis from the second half of 2010 through 2014 and monthly thereafter. The semiannual gas production changes the units of observation from quarterly changes on labor market outcomes to semiannual changes in first-difference and difference-in-differences estimation. Monthly county unemployment rates and the county population in 2010 are appended to the data.

Shale gas production is not uniformly distributed over space but concentrated within regions. As our interest is in the relative impact of gas production compared to similar counties, we restrict our analysis to rural counties as they were classified by the Bureau of the Census in 2010. Classifications are not without debate as some rural counties are parts of larger Metropolitan Statistical Areas (MSA) where residents frequently commute to central business districts. In Pennsylvania, Butler County is part of the Pittsburgh MSA and it is a commutable distance from downtown Pittsburgh and nearby employment centers. The growth in townships bordering on the larger Allegheny County since 2010 exceeded the population growth in the entire county. Since these gains reflect economic opportunities in Pittsburgh and other communities in Allegheny County, we believe that it is more accurately regarded as an urban county and we drop it from the group of rural counties.[4] We categorize counties by the quantity of gas production into five groups based on the quartiles of the counties producing natural gas. Gas counties are numbered from zero to four, with zero producing no gas and four the top 25 percent of gas producing counties where most of the gas is produced.

Figure 9.1 reports employment in all rural counties by quartiles of shale gas production, no gas (=0), gas production in the lowest quartile (=1), gas production in the second quartile (=2), gas production in the third quartile (=3), and high-gas counties (=4). The figure shows a general inverse relationship between

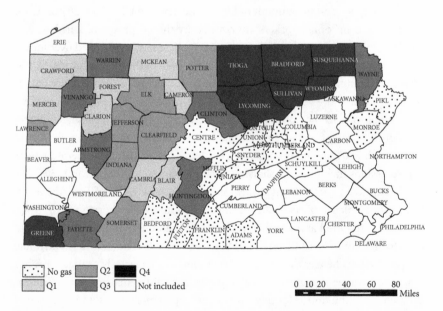

FIGURE 9.1. Rural counties by gas production quartiles. Compiled by authors.

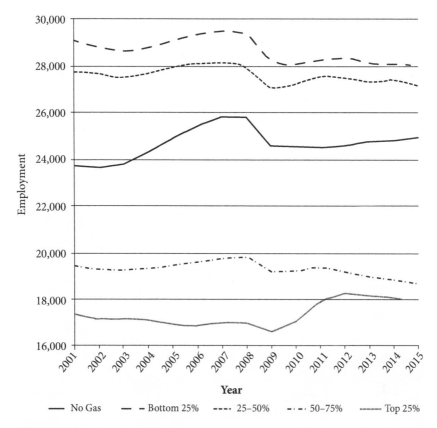

FIGURE 9.2. Annual employment by gas county quartile. Compiled by authors.

employment and gas production with one exception. While the two highest gas producing counties have the lowest employment, counties with gas employ fewer than lower gas producing counties. The counties with the most employment, which were in the lowest quartile of gas production, employed about two-thirds more than in the highest gas producing counties.

Figure 9.2 charts the growth of employment over time by gas production categories. High-gas counties did not participate in the economic expansion from 2001 to 2007 as employment fell by nearly 3 percent from 2001 to 2006 and then by another 1.5 percent from 2006 to 2009. Employment grew in all other categories from 2001 to 2006, but only counties with no gas employed more workers in 2009 than in 2001. The advent of shale gas production in 2010 coincided with a sharp increase in employment in high-gas counties that was unmatched by others as employment grew by 9.5 percent from 2009 to 2012, which more than offset the earlier losses. The drop in jobs after 2012 reflects reduced exploration as production continued to increase. Among gas producing counties, only high-gas counties enjoyed job growth over the entire period.

The change in average weekly wages by gas production, reported in figure 9.3, is even more striking than the employment change. Workers in counties with no gas, the most gas, and in the second quartile of gas production enjoyed the highest earnings at the start of the period. Real wages were stagnant in high-gas counties through 2008 compared to small gains elsewhere. After 2008, real wages grew sharply in high-gas counties and ended the period 17 percent higher than in 2008, which was more than twice the wage growth in other county groups.

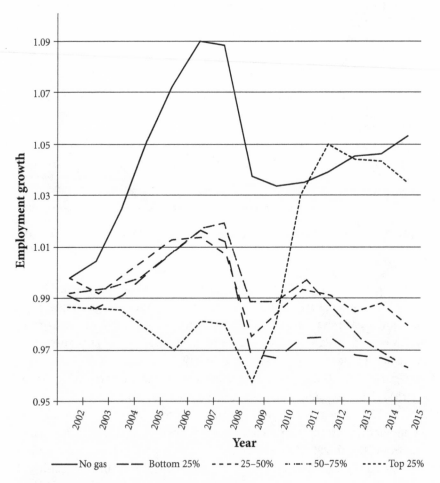

FIGURE 9.3.   Annual employment growth by gas county quartile. Compiled by authors.

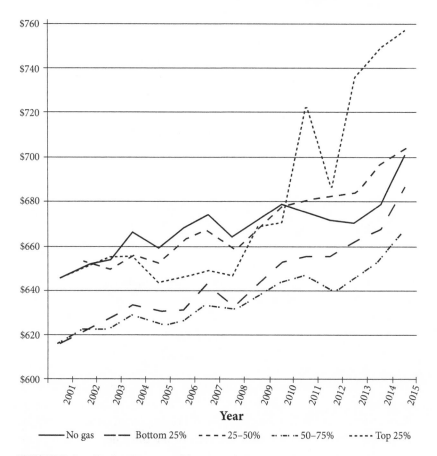

FIGURE 9.4. Real average weekly wages by gas county quartile. Compiled by authors.

Figure 9.4 reports the growth in real average weekly wages and shows that the decline in real wages in 2005 and 2006 in high-gas counties was dramatically reversed at the time that shale gas production began. The 17 percent earnings gain in high-gas counties is over twice the growth in no-gas counties and 6 percent more than in the next highest group. Unlike employment, earnings grew in high-gas counties since 2012 as local economies benefit from continued production. Hence, shale gas production appears to have had a significant impact on the well-being of workers in counties that had previously been losing jobs for many years.

The timing of the impact of gas exploration and development on local economic activity is unclear. The exploration and development require more resources than the extracting of the gas from the wells. As noted, we use three measures of gas production. The first is an indicator or binary variable for counties in the top quartile of gas production. Another is the semiannual change in gas production within each county and the last is the total amount of shale gas produced over the entire period, which captures development activity along with gas production. We also use three measures of economic activity all in natural logarithm—employment change, changes in total wages, and changes in average weekly wages. We begin with a first-difference estimator, where the dependent variable corresponds to semiannual changes in local labor market outcomes. Specifically, for employment we estimate

$$(1) Y_{it} = ln\bar{Y}_t - ln\bar{Y}_{t-1}$$

with $\quad \bar{Y}_t = (Y_t + Y_{t-1}) \div 2 \qquad$ and $\bar{Y}_{t-2} = (Y_{t-2} + Y_{t-3}) \div 2$

For example, if $\bar{Y}_t$ is the average for quarters 3 and 4, then $\bar{Y}_{t-2}$ is the average for quarters 1 and 2 of the same year. The changes in the average real total wages and the real average weekly wages are similarly defined. Defining $y_t = Y_t - Y_{t-2}$ we estimate:

$$(2) \qquad y_{it} = \alpha_0 + \alpha_1 G_{it} + \alpha_2 UR_{it} + \alpha_3 \bar{X}_{it} + \varepsilon_{it}$$

where G_{it} is the gas production in county i and quarter t, UR_{it} is the unemployment rate in county i and quarter t, and $\bar{X}_{it}$ includes county population and an indicator for whether or not the county is in a metropolitan statistical area. We estimate the first-difference models for each of the three dependent variables for five samples. The first includes employment aggregated over all industries in a county in each time period. The next restricts the sample to only private-sector employment, the third selects only the nontraded construction sector, the fourth the manufacturing sector, and the last the retail trade and service sector. The theoretical prediction of Corden and Neary (1982) is that the stimulus from the resource boom is stronger in the nontraded sectors including construction and retail trade and services. We also estimate a difference-in-differences model (DD). Here the treatment is a high-gas county, and the treatment occurs with the increase in Marcellus unconventional gas production in the second half of 2010. The DD model includes an indicator variable for the treatment group (i.e., high-gas county), an indicator for the treatment period from the second half of 2010, and an interaction term which shows how the outcome of interest differs between the treatment and control (other) counties. County unemployment rates and an indicator if the rural county was part of a metropolitan statistical area (MSA) are also in the DD model.

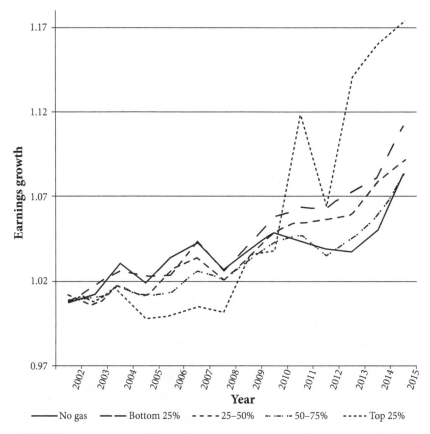

FIGURE 9.5.   Annual real wage growth by gas county quartile. Compiled by authors.

Findings

Table 9.1 reports the first-difference estimates of the semiannual changes in the three outcomes. Col. 1 of panel A reports that employment grew in high-gas counties over the period. The precisely estimated coefficient shows a 0.7 percent semiannual increase compared to all other rural counties, which results in a 5 percent gain over the five-and-a-half-year study period. The economic stimulus of shale gas is larger in the private sector as the estimate in the sample of private-sector workers in col. 2 is slightly larger, albeit less precisely estimated than in col. 1. The next three columns find support for "Dutch Disease"[5] as employment grew in construction and retail trade in high-gas counties. The largest gains were estimated in construction as employment grew by 4.6 percent semiannually.

The coefficients in construction and retail trade were significant at the 10 percent and 1 percent levels respectively. Employment in the traded manufacturing sector fell in high-gas counties, although the impact was not precisely estimated. Substituting the change in gas production for a high-gas county indicator reduces the impact of gas production as only the coefficient in retail trade reaches conventional levels of statistical significance. This is not surprising as the resources required to produce the gas are small compared to those needed for exploration and drilling. It is also unclear how the timing of royalty checks to local landowners compares to gas production. Panel C includes the gas production over the entire period that exploits the resources needed for all phases of production from exploration to extracting the gas from the well. Employment grows more with gas production, and the impact is precisely estimated in all samples except for manufacturing. The coefficients are significant at the 1 percent level in the other samples, and the 0.1 percent estimate in col. 1 and col. 2 combined with gas production implies that employment in high-gas counties grew by about 0.33 percent more semiannually than employment in the third quartile and about 0.6 percent more semiannually than in the lowest quartile. Gas production had a particularly large (0.18 percent) effect on retail trade employment as the effect includes both the employment stimulus of gas production and royalty payments to landowners.

Table 9.2 reports how the ln of total wages responds to different measures of gas production. Panel A reports that total wages are high in high-gas counties in all samples except for manufacturing; however only the estimate in the sample that excludes the public sector is statistically significant and then only at the 10 percent level. Omitting the public sector doubles the wage growth in high-gas counties as the private sector may be more responsive to short-run changes in demand conditions brought about by gas production. The nearly 0.2 percent semiannual earnings gain after the beginning of shale gas production represents a clear departure from the stagnant wage growth between 2001 and 2009. Total wages were not responsive to changes in the ln of gas production as the impact was not precisely estimated in any sample as listed in panel B. The estimate was negative in the private sector and in the manufacturing sector and only the coefficient for all groups in col. 1 was larger than its standard error. The ln of total gas production leads to an increase in semiannual total wage growth in all samples except manufacturing, although again only the estimate for the private sector reached the 10 percent level of significance. The impact of total gas production on the ln of total wage growth is twice as large in the private sector as it is when the public sector is included.

Average weekly wages were not responsive to any measure of shale gas production as its impact was not significant in any sample shown in table 9.3. Panel

TABLE 9.1 Estimates of ln employment change

	ALL	PRIVATE	CONSTRUCTION	MFG	RETAIL
PANEL A		REGRESSION ESTIMATES OF LN EMPLOYMENT CHANGE			
Top 25% gas	0.0069	0.0075	0.0459	−0.0054	0.0095
counties	(0.0033)**	(0.0042)*	(0.0264)*	(0.0065)	(0.0041)**
$UR_t - UR_{t-1}$	0.0172	0.0317	0.0472	0.0084	0.0335
	(0.0122)	0.0152	(0.0706)	0.0318	0.0183
MSA	−0.0017	−0.0024	−0.0251	0.0018	−0.0098
	0.0024	0.0029	0.0198	0.0043	0.0065
Population	3.67E-05	3.16E-05	2.73E-04	1.66E-05	1.08E-04
(0000)	2.64E-08***	3.23E-08***	2.51E-07***	7.43E-08***	7.11E-08***
Quarter 2	−0.0192	−0.0383	−0.2526	−0.0209	−0.045
	0.0020	0.0025	0.0136	0.0049	0.0029
Constant	0.0098	0.0249	0.1171	0.0099	0.0215
R-squared	0.2187	0.4022	0.5150	0.0598	0.4390
PANEL B		GAS PRODUCTION			
Gast -Gast-1	0.0005	0.0002	0.0006	0.0012	0.0023
	0.0004	0.0006	0.0026	0.0015	0.0012
$UR_t - UR_{t-1}$	−0.0354	−0.0356	0.0665	−0.0493	0.0173
	0.0154	0.0205	0.1220	0.0406	0.0363
MSA	−0.0002	0.0001	0.0086	0.0033	−0.0137
	(0.0026)***	(0.0032)***	(0.0251)***	(0.0045)***	(0.0084)***
Population	1.98E-05	5.48E-06	−4.05E-04	−1.64E-05	1.39E-04
(0000)	3.10E-08***	3.95E-08***	3.51E-07***	8.06E-08***	8.82E-08***
Quarter 2	−0.0112	−0.0274	−0.2473	−0.0144	−0.042
	0.0023	0.0030	0.0181	0.0064	0.0041
Constant	−0.0058	0.0048	0.1754	−0.0092	0.0165
R-squared	0.1255	0.2974	0.4788	0.0471	0.3747
PANEL C		TOTAL GAS PRODUCTION			
Total Gas	0.0011	0.0011	0.0005	−0.0005	0.0018
	0.0004	0.0005	1.9900	0.0011	0.0006
$UR_t - UR_{t-1}$	0.0173	0.0322	0.0160	−0.0037	0.0331
	0.0128	0.0160	2.0200	0.0341	0.0199
MSA	0.0011	0.0005	0.0030	0.0019	−0.0083
	0.0025	0.0030	0.1700	0.0041	0.0064
Population	2.87E-05	1.79E-05	3.76E-05	2.46E-05	1.30E-04
(0000)	3.03E-08***	3.76E-08***	4.80E-01***	8.16E-08***	7.56E-08***

(Continued)

TABLE 9.1 Continued

	ALL	PRIVATE	CONSTRUCTION	MFG	RETAIL
PANEL C			TOTAL GAS PRODUCTION		
Quarter 2	−0.0181	−0.0369	0.0028	−0.021	−0.0459
	0.0022	0.0028	(13.2500)	0.0053	0.0032
Constant	−0.0092	0.0068	0.0103	0.0131	−0.0105
R-squared	0.1939	0.3736	0.5317	0.0577	0.4412

*Significant at the 10% level
**Significant at the 5% level
***Significant at the 1% level

TABLE 9.2 Estimates of ln wage change

	ALL	PRIVATE	CONSTRUCTION	MFG	RETAIL
PANEL A			REGRESSION ESTIMATES OF LN WAGE CHANGE		
Top 25% gas	0.0094	0.018	0.0551	−0.0013	0.0125
counties	(0.0178)	0.0102	(0.0454)	(0.0107)	(0.0106)
$UR_t - UR_{t-1}$	0.0862	0.1358	0.1245	0.0944	0.0503
	(0.0652)	(0.0330)***	0.1289	0.0506	0.0608
MSA	−0.001	−0.0054	−0.0085	−0.0009	−0.0107
	(0.0058)***	(0.0079)***	(0.0425)***	(0.0111)***	(0.0136)***
Population	3.05E−05	1.56E−04	2.48E−04	1.09E−04	1.53E−04
(0000)	1.10E-07***	1.38E-07***	5.23E-07***	1.66E-07***	2.00E-07***
Quarter 2	−0.0719	−0.1249	−0.4592	−0.1059	−0.1384
	(0.0091)***	(0.0072)***	(0.0237)***	(0.0086)***	(0.0078)***
Constant	0.058	0.0849	0.2371	0.0683	0.07
R-squared	0.1627	0.4902	0.5274	0.3331	0.4895
PANEL B			GAS PRODUCTION		
Top 25% gas	0.0041	−0.0002	0.0031	−0.0006	0.0034
counties	(0.0038)	(0.0014)	(0.005)	(0.0019)	(0.0031)
$UR_t - UR_{t-1}$	0.0028	−0.0855	0.0566	−0.1991	−0.1283
	(0.0529)	0.0379	(0.2161)	(0.0617)***	(0.1285)
MSA	0.0038	0.002	0.0376	0.0084	−0.0157
	(0.0068)	(0.0085)	(0.0531)	(0.0103)	(0.0165)
Population	−1.18E−04	1.02E−04	−6.55E−04	−1.00E−05	1.73E−04
(0000)	1.17E-07***	1.62E-07***	7.00E-07***	1.58E-07***	2.28E-07***
Quarter 2	−0.054	−0.0996	−0.4396	−0.0738	−0.1254
	(0.0069)***	(0.0089)***	(0.0317)***	(0.0101)***	(0.0118)***
Constant	0.0386	0.0232	0.2853	−0.0171	0.0224
R-squared	0.2178	0.3502	0.4802	0.2504	0.4054

	ALL	PRIVATE	CONSTRUCTION	MFG	RETAIL
PANEL C		LN TOTAL COUNTY GAS PRODUCTION			
Top 25% gas	0.0017	0.0032	0.0043	−0.0001	0.0024
counties	(0.0013)	0.0017	(0.0056)	(0.0019)	(0.0019)
UR_t -UR_{t-1}	0.0863	0.1357	0.1581	0.0843	0.0442
	(0.0697)	(0.0345)***	(0.1321)	(0.0537)	(0.0655)
MSA	0.0034	0.001	0.0145	0.0047	−0.007
	(0.0067)	(0.0076)	(0.0446)	(0.0097)	(0.0129)
Population	1.25E-05	1.59E-04	−1.86E-04	4.18E-05	1.43E-04
(0000)	1.67E-07***	1.62E-07***	6.21E-07***	1.62E-07***	2.14E-07***
Quarter 2	−0.0682	−0.1228	−0.4508	−0.1052	−0.141
	(0.0103)***	(0.0081)***	(0.0248)***	(0.0093)***	(0.0087)***
Constant	0.0277	0.0265	0.2091	0.0686	0.0299
R-squared	0.1374	0.4629	0.5312	0.3280	0.4772

A shows that employment was relatively higher in high-gas counties, but only the impact in the private sector has an estimate that exceeded its standard error. The change in gas production had little impact on average wages, and the estimate was negative for all private-sector workers and in the manufacturing sector. The results in panel C show a stronger impact of total gas production on average weekly wages, but again the estimate was only larger than its standard error for all private-sector workers. While gas production increases employment and total wages, it has little effect on average weekly wages. Additional employment owing to gas production especially in retail trade could be in low-wage jobs common for new entrants, which depresses average wages but increases the total wage bill.

Table 9.4 panel A presents difference-in-differences estimates for employment change in all five samples. As there was no shale gas production during the first nine years of the study period, the interest is in how employment changed in gas counties in the last five and a half years when gas was extracted. The interaction term between high-gas counties and the period in which gas was produced, Top 25 Percent Gas Counties (High Gas Post 2009), shows the differences in employment growth in high-gas counties and other counties when gas began to flow. Col. 1 reports that in the sample of all sectors, employment increases 0.65 percent semiannually or 1.3 percent per year compared to other counties, and the estimate is significant at the 10 percent level. Employment grew by about 1.65 percent annually in the private sector suggesting that the government employment responds more slowly to the stimulus of gas production; however the coefficient is just below the 10 percent level. Employment in both the traded

TABLE 9.3 Estimates of ln average wage change

	ALL	PRIVATE	CONSTRUCTION	MFG	RETAIL
PANEL A		REGRESSION ESTIMATES OF LN AVERAGE WAGE CHANGE			
Top 25% gas counties	0.0026	0.0096	0.0124	0.0021	0.0028
	(0.0182)	(0.0074)	(0.0261)	(0.0078)	(0.0100)
UR_t -UR_{t-1}	0.0697	0.1059	0.1004	0.0926	0.0159
	(0.0668)	(0.0257)***	(0.0934)	(0.0437)***	(0.0609)
MSA	0.0006	−0.0025	0.0125	−0.0026	−0.0008
	(0.0051)	(0.0067)	(0.0312)	(0.0094)	(0.0101)
Population	−8.05E-06	1.04E-04	−1.14E-05	1.05E-04	4.01E-05
(0000)	1.12E-07***	1.19E-07***	3.83E-07***	1.31E-07***	1.79E-07***
Quarter 2	−0.0522	−0.0863	−0.2189	−0.0847	−0.0931
	(0.0093)***	(0.0057)***	(0.0144)***	(0.0068)***	(0.0075)***
Constant	0.0484	0.0621	0.1313	0.0587	0.0486
R-squared	0.0909	0.4256	0.4187	0.4377	0.3214
PANEL B		CHANGE IN GAS PRODUCTION			
Top 25% gas counties	0.004	−0.0004	0.0022	−0.0015	0.0012
	(0.0041)	(0.0012)	(0.0031)	(0.0015)	(0.0029)
UR_t -UR_{t-1}	0.0438	−0.0478	−0.0052	−0.1428	−0.155
	(0.0567)	(0.0324)	(0.1623)	(0.0479)***	(0.1305)
MSA	0.0041	0.0021	0.0321	0.0047	−0.002
	(0.006)	(0.0073)	(0.0406)	(0.0089)	(0.0116)
Population	−1.50E-07	7.44E-08	−3.22E-07	3.32E-08	3.09E-08
(0000)	1.20E-07***	1.37E-07***	5.21E-07***	1.10E-07***	2.02E-07***
Quarter 2	−0.0421	−0.0714	−0.2007	−0.0587	−0.0824
	(0.0068)***	(0.0071)***	(0.0193)***	(0.0071)***	(0.0119)***
Constant	0.0464	0.0205	0.1205	−0.0091	0.0035
R-squared	0.1523	0.2976	0.3565	0.3041	0.2541
PANEL C		MAXIMUM LN GAS PRODUCTION			
Top 25% gas counties	0.0006	0.002	0.0009	0.0003	0.0005
	(0.0012)	(0.0014)	(0.0035)	(0.0013)	(0.0019)
UR_t -UR_{t-1}	0.0692	0.1055	0.1145	0.0948	0.0104
	(0.072)	(0.0272)***	(0.0992)	(0.0452)***	(0.0655)
MSA	0.0023	0.0007	0.0215	0.0026	0.0013
	(0.006)	(0.0063)	(0.0331)	(0.0082)	(0.0092)
Population	−1.68E-08	1.18E-07	−1.82E-07	3.55E-08	7.76E-09
(0000)	1.72E-07***	1.37E-07***	4.80E-07***	1.22E-07***	1.92E-07***
Quarter 2	−0.0494	−0.0853	−0.2097	−0.0838	−0.0948
	(0.0105)***	(0.0064)***	(0.0159)***	(0.0068)***	(0.0083)***
Constant	0.0368	0.0257	0.1283	0.0565	0.0414
R-squared	0.0753	0.4059	0.3955	0.3684	0.3102

and nontraded industries grew with the production of gas although no coefficient is precisely estimated. Total wages also grew in high-gas counties during the gas boom according to the results presented in panel B. Total wages grew in all five samples, and again the coefficient in the sample of private-sector workers was significant at the 5 percent level. As with employment, total wages grew about twice as fast in the private sector compared to the sample of both private- and public-sector employment. Total wages grew in each of the other sectors, and the impacts in both construction and retail trade were closer to the border of statistical significance than in manufacturing. Unlike in the first-difference estimation, average weekly wages also increased in high-gas counties with the production of gas and again the impact in the private sector is precisely estimated. Unlike the other measures of labor market performance, the increase in average weekly wages in the sample of all industries reported the lowest t-value.

TABLE 9.4 Difference-in-differences estimation

	ALL	PRIVATE	CONSTRUCTION	MFG	RETAIL
PANEL A	DEPENDENT VARIABLE: LN EMPLOYMENT CHANGE				
Top 25% gas	−0.0002	−0.0003	0.0082	−0.0096	0.003
counties	(0.0023)	(0.0031)	(0.016)	0.0056	(0.0047)
High gas	0.0065	0.0083	0.046	0.0018	0.0065
post-2009	0.0038	(0.0053)	(0.0331)	(0.0078)	(0.0061)
UR_t-UR_{t-1}	−0.0248	−0.0381	−0.1409	−0.0693	−0.011
	(0.0047)***	(0.0064)***	(0.0239)***	(0.0111)***	(0.0069)
MSA	−0.0007	−0.001	−0.0124	0.0033	−0.0009
	(0.0014)	(0.0244)	(0.0107)	(0.0029)	(0.003)
Constant	0.0103	0.0244	0.1847	−0.0123	0.0106
R-squared	0.1125	0.1038	0.0546	0.2174	0.0221
PANEL B	DEPENDENT VARIABLE: LN TOTAL WAGE CHANGE				
Top 25% gas	−0.0026	−0.0102	0.0031	−0.0099	−0.0011
counties	(0.0042)	(0.009)	(0.0228)	(0.0091)	(0.0078)
High gas	0.0175	0.0326	0.0719	0.0148	0.0213
post-2009	(0.0199)	(0.0135)***	(0.0552)	(0.0132)	(0.0133)
UR_t-UR_{t-1}	−0.0360	−0.0688	−0.2147	−0.1172	−0.0308
	(0.0099)***	(0.0151)***	(0.0440)***	(0.0188)***	0.0119
MSA	−0.0005	0.0003	−0.011	0.0056	−0.0005
	(0.0042)	(0.0051)	(0.0178)	(0.0055)	(0.006)
Constant	0.0289	0.0653	0.3028	0.0244	0.0584
R-squared	0.033	0.0396	0.0401	0.0771	0.042

(Continued)

TABLE 9.4 Continued

	ALL	PRIVATE	CONSTRUCTION	MFG	RETAIL
PANEL C	DEPENDENT VARIABLE: LN AVERAGE WEEKLY WAGE CHANGE				
Top 25% gas	–0.0024	–0.01	–0.0064	–0.0014	–0.0042
counties	(0.003)	(0.007)	(0.0105)	(0.0072)	(0.0058)
High gas	0.0116	0.0241	0.0337	0.0117	0.0148
post-2009	(0.0203)	(0.0096)***	(0.0286)	(0.0106)	(0.0107)
UR_t -UR_{t-1}	–0.0108	–0.0328	–0.0749	–0.0498	–0.0198
	(0.0074)	(0.0116)***	(0.0277)***	(0.0153)***	(0.0088)***
MSA	0.0001	0.001	0.0009	0.0018	0.0004
	(0.0035)	(0.0036)	(0.0099)	(0.0043)	(0.0042)
Constant	0.0184	0.0422	0.1267	0.0383	0.0479
R-squared	0.0252	0.0306	0.0249	0.0437	0.0547

Inferences

Rural counties with the most shale gas enjoyed employment and wage gains following the initial production of gas in 2009. The impact is somewhat stronger on employment as our estimates imply that employment grew by about 3.5 percent in counties in the upper quartile of gas production compared to all others. We also find that the stimulus from gas production follows the theoretical prediction of Corden and Neary (1982) as there is a significant stimulus in the nontraded construction and retail trade industries and no effect on manufacturing that produces goods that are generally consumed elsewhere. We also find that the contemporaneous semiannual change in shale gas production had little effect on employment or wages, but the county variation in production over the entire period had a positive and significant effect on employment in all the samples we consider with the exception of manufacturing, and a positive impact on total real wage for workers in all private-sector employment. Real average weekly wages were not responsive to any measure of shale gas production, which could be due to additional employment at the lower end of the wage scale.

The impact of gas production is likely stronger than our estimate shows as we often compare outcomes in the top quartile to a comparison group that includes counties with no gas production and counties in other quartiles of gas production. Furthermore, since many of these counties share borders, the impact of gas production in one county likely has spillover effects on others.

The 3.5 percent employment gain translates into roughly 850 jobs added in high-gas counties over the nearly six-year study period. While this is about two-thirds of what is estimated by Weber (2012), his sample covers a much larger region. We have no measure of the costs associated with fracking or other costs of shale gas production, which precludes a cost-benefit calculation. We do find that the economic benefits of gas production are accruing to places that have been losing jobs for many years and suffer higher unemployment rates than other areas, and that much of the concern about the environmental cost of shale gas is echoing from urban areas far removed from these rural counties. While it is clearly possible that the costs of shale gas production far outweigh the benefits, we do find a substantial benefit from shale gas—a finding that ought to have a role in the debate about shale gas development.

Notes

1. These counties are Bradford, Lycoming, McKean, Sullivan, Susquehanna, Tioga, Warren, and Wyoming.

2. Excluded are members of the armed forces, the self-employed, proprietors, domestic workers, unpaid family workers, and railroad workers covered by the railroad unemployment insurance system.

3. Only 112 of 7,772 industry-county employment pairs are missing due to disclosure rules from 2001 through 2010.

4. Dropping Butler County affects the size of the rural sector charted in figures 9.1 and 9.2, but it has little change on the regression estimates.

5. The concept developed by Corden and Neary (1982) refers to the impact that a boom in one sector has on other industries. Goods produced in nontraded industries such as construction and retail trade benefit from the additional revenue from boom industries. Traded-goods industries suffer as their costs increase due to competition for resources in the boom sector.

References

Black, Dan, Terra McKinnish, and Seth Sanders. 2005. "The Economic Impact of the Coal Boom and Bust." *Economic Journal* 115:449–76.

Considine, Timothy J., Robert Watson, and Seth Blumsack. 2010. "The Economic Impacts of the Pennsylvania Marcellus Shale Natural Gas Play: An Update." Penn State University Department of Energy and Mineral Engineering. https://www.researchgate.net/publication/228367795_The_economic_impacts_of_the_Pennsylvania_Marcellus_Shale_natural_gas_play_An_update.

Corden, W. Max, and J. Peter Neary. 1982. "Booming Sector and De-Industrialization in a Small Open Economy." *Economic Journal* 92 (368): 825–48.

Jacobsen, Grant D., and Dominic P. Parker. 2016. "The Economic Aftermath of Resource Booms: Evidence from Boomtowns in the American West." *Economic Journal* 126:1092–1127.

Jacquet, Jeffrey. 2006. "A Brief History of Drilling 1995–2005." Wyoming: Socioeconomic Analyst Advisory Committee. Accessed January 12, 2013. http://www.sublettewyo.com/DocumentCenter/Home/View/346.

James, Alex, and David Aadland. 2011. "The Curse of Natural Resources: An Empirical Investigation of U.S. Counties." *Resource and Energy Economics* 33 (2): 440–53.

Marchand, Joseph. 2012. "Local Labour Market Impacts of Energy Boom-Bust-Boom in Western Canada." *Journal of Urban Economics* 71 (1): 165–74.

Michaels, Guy. 2010. "The Long Term Consequences of Resource-Based Specialisation." *Economic Journal* 121:31–57.

Papyrakis, Elissaios, and Reyer Gerlagh. 2007. "Resource Abundance and Economic Growth in the United States." *European Economic Review* 51 (4): 1011–39.

Weber, Jeremy G. 2012. "The Effects of a Natural Gas Boom on Employment and Income in Colorado, Texas, and Wyoming." *Energy Economics* 34:1580–88.

SHALE ENERGY AND REGIONAL ECONOMIC DEVELOPMENT IMPACTS IN NORTHWEST PENNSYLVANIA

Erik R. Pages, Martin Romitti, and Mark C. White

If you read the business press in the years before the COVID-19 pandemic, you would have an optimistic feeling about the state of manufacturing in the United States. It seems as if a new report touting US manufacturing renaissance comes out nearly every day from business consultants, academic researchers, and government agencies. Support for manufacturing was a major campaign plank for Donald Trump, who subsequently promised to "make manufacturing great again." This support continued during Trump's administration, as did his commitment to reopen the manufacturing economy as the country attempts to "flatten the curve" of the pandemic that started in 2020.

There are many valid reasons for the pre-COVID optimism. Operating costs in the United States are declining in comparison to other global competitors. The combination of cheaper energy in the United States and rising costs and uncertainty in overseas locations, such as China, enhances the competitiveness of US locations as sites for new or expanded manufacturing facilities and operations. At the same time, many US manufacturers have embraced the latest management thinking and practices, becoming global leaders in terms of innovation, efficiency, and productivity. Nonetheless, these promising trends are no guarantee of long-term success or prosperity. US manufacturers must recover from years of challenges and downsizing. This requires a more skilled and talented workforce, a rebuilding of the "industrial commons." Along the way, US manufacturers—large, medium, and small firms—must commit to developing more resilient and competitive supply chains. Smaller firms and new market entrants need to showcase their capabilities, and larger firms and buyers must be able to easily access qualified US supply chain partners that possess the capabilities to meet production requirements and time-to-market guidelines.

Industry groups and manufacturing advocates have devoted substantial effort to nurturing such an industrial commons, built around robust supply chains and reshoring of work to US locations. A US National Institute of Standards and Technology-Manufacturing Extension Partnership (NIST-MEP)–backed study offers strong evidence that effective supply chain management remains a big challenge for US manufacturers (GENEDGE Alliance 2012). The analysis surveyed manufacturers to identify some of their critical supply chain related needs. A number of key challenges were identified, with top priority given to the following items:

- Companies suffer from a lack of collaboration and visibility within supply chains.
- They have a limited understanding of the total cost of ownership.
- They lack an overall synchronized plan for supply chain management.

These factors have helped trigger growing interest in what many experts refer to as supply chain optimization, a process whereby firms develop a supply chain strategy that is closely aligned with overall corporate strategy and which helps to improve the speed of delivery while also reducing operational costs.

This interest occurs at the local, state, and federal levels. Across the United States, many regions and states are engaged in new initiatives tied to reshoring and supply chain optimization. In Pennsylvania, the statewide PA Made Network, led by the state's Industrial Resource Center Network, is promoting a host of efforts to support reshoring initiatives. At the federal level, the Obama administration supported and promoted the Investing in Manufacturing Communities Partnership (IMCP), a multiagency series of initiatives designed to support the revitalization of manufacturing across the United States. This effort included several NIST-MEP programs focused on supply chain optimization. Under the Trump administration, new policies to stimulate manufacturing focus on regulatory relief, tax reductions, and tougher trade enforcement.

As part of the IMCP initiative, the authors developed a supply chain analysis for northwest Pennsylvania. The region is an ideal candidate for this type of analysis. It is home to a large base of manufacturing firms in a diverse mix of sectors. The region's various business support and economic development agencies have a long history of collaboration and partnership, and thus are well positioned to develop new and improved programs to help strengthen regional manufacturing supply chains.

In the case of shale gas related manufacturing, northwest Pennsylvania sits in an advantageous position at the juncture of the Marcellus and Utica Shale formations, which are among the world's richest sources of shale gas. The shale gas and oil revolution has been considered to be a "game changer" that may help to transform US manufacturing (Gray, Linn, and Morgenstern 2018; Houser and Mohan 2014). The analysis presented in this chapter assesses the local potential to benefit from proximity to shale gas resources. Some of these opportunities may be directly related to drilling and extraction, but long-term and more promising prospects are likely to be found in midstream activities, such as transportation, storage, and distribution, and a diverse mix of downstream activities. These downstream markets are less well understood but offer massive potential. Similar large-scale impacts are expected across other energy-intensive manufacturing sectors, many of which are concentrated in northwest Pennsylvania. As the economy reopens after the current pandemic, local and state governments are increasingly likely to search out such opportunities, making such analysis especially timely going forward.

This chapter begins with a general introduction to the northwest Pennsylvania economy with a focus on the region's manufacturing assets, challenges, and opportunities. The report next provides in-depth analyses of the regional manufacturing supply chains related to shale gas development—with a basic introduction to the industry and its operations in northwest Pennsylvania. We then develop a regional value chain analysis and conclude with a review of key issues and regional economic opportunities.

The Northwest Pennsylvania Economy

This research focuses on an eight-county region of northwest Pennsylvania, including the counties of Clarion, Crawford, Erie, Forest, Lawrence, Mercer, Venango, and Warren (see figure 10.1). The communities have an extensive track record of regional cooperation through a number of alliances, including the Northwest Pennsylvania Regional Planning and Development Commission (Northwest Commission), the Local Development District, and Partnership for Regional Economic Performance (PREP) Region Northwest, a state economic development partnership.

Regional economic development projects have focused on small business development and manufacturing, two key drivers of the regional economy. According to the US Census Bureau, approximately 70 percent of local establishments in the region have fewer than ten employees, with many in manufacturing sectors, as northwest Pennsylvania has been a major national and global industrial center since the late 1800s. In addition to major manufacturing activities, the region was the birthplace of the modern oil industry, home to Edwin Drake's first oil well drilled near Titusville, in Crawford County, in 1859.

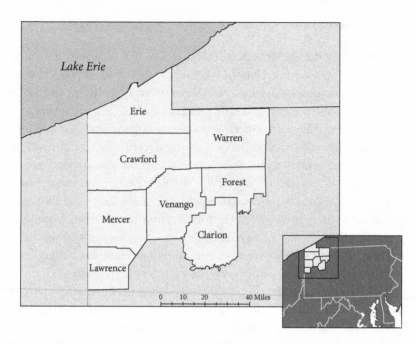

FIGURE 10.1. Project region in northwest Pennsylvania. Compiled by authors.

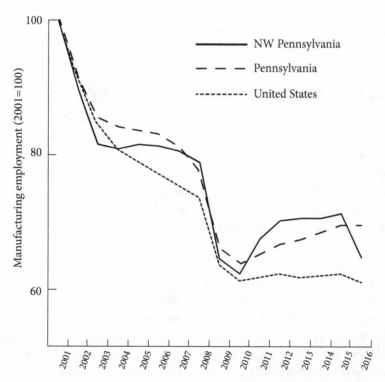

FIGURE 10.2. Manufacturing jobs performance, 2001–2016. Compiled by authors.

This manufacturing legacy lives on today. In recent years, manufacturing jobs have declined in northwest Pennsylvania at a rate similar to statewide and national levels, but the overall concentration of regional manufacturing activity remains relatively high. Manufacturing jobs account for almost 16 percent of employment in the eight-county region, as opposed to 9 percent for the Commonwealth of Pennsylvania (see figure 10.2). According to data from Economic Modeling Specialists International (2017), manufacturing is the region's second largest sector (behind health care), and manufacturing activity accounts for 21 percent of gross regional product. Not surprisingly, manufacturers represent many of the region's larger employers. Even though manufacturers account for only 7.7 percent of total establishments, they represent 31.3 percent of regional establishments employing a hundred or more employees.

The Northwest Pennsylvania Shale Gas Supply Chain

The emergence of shale gas extraction in the Marcellus and Utica Shale basins is one of the most exciting—and controversial—economic forces now driving Pennsylvania's economy. Thanks to the development of shale gas, Pennsylvania ranks among the top states in oil and gas related development. Between 2009 and 2012, the commonwealth saw a nearly 260 percent jump in oil and gas related jobs, placing Pennsylvania second in the United States for such job growth (Cruz, Smith, and Stanley 2014). In 2016, Pennsylvania had almost 18,900 oil and gas related jobs making it the sixth largest state for these activities, behind more traditional oil and gas states like Texas, Oklahoma, and Louisiana, as well as California and Colorado. Oil and gas related employment is down significantly since it peaked in 2014, and nationally these industries lost about a quarter of total employment between 2014 and 2016. In spite of the slowdown in activity around 2016, oil and gas related employment in Pennsylvania remained twice as large as it was before drilling began in the Marcellus in 2001, and in northwest Pennsylvania it is more than 50 percent greater.

To date, shale gas investments in Pennsylvania have been centered in the north central and southwest parts of the state. Northwest Pennsylvania has seen more limited shale gas activity, although it supports a small but relatively stable oil industry. Overall, nearly 1,400 people work directly in the oil and gas industries in the northwest Pennsylvania region (see table 10.1). Later, we show the individual industries that have been driving this employment growth. Actual crude petroleum and natural gas extraction employs nearly 330 jobs, but support

TABLE 10.1   Employment trends in key shale-related industries

INDUSTRY	NORTHWEST PENNSYLVANIA				PENNSYLVANIA			
	2006	2016	CHANGE (06–16)	ANNUAL GROWTH RATE (06–16)	2006	2016	CHANGE (06–16)	ANNUAL GROWTH RATE (06–16)
Crude petroleum extraction	250	268	18	0.7%	821	2,193	1,372	16.7%
Natural gas extraction	258	224	–34	–1.3%	1,220	3,267	2,047	16.8%
Drilling oil and gas wells	133	73	–60	–4.5%	1,178	1,224	46	0.4%
Support activities for oil and gas operations	215	390	175	8.1%	1,809	7,399	5,590	30.9%
Natural gas distribution	414	431	17	0.4%	3,899	4,787	888	2.3%
Totals	1,271	1,385	114	0.9%	8,927	18,870	9,943	11.1%

Source: Author calculations from data from Economic Modeling Specialists International (2017).

activity for oil and gas activities is both the largest and fastest growing industry since 2001. This latter sector's growth has been driven by shale gas related activities.

The recent downturn in shale gas activity has generated job losses in many key sectors, but future prospects look brighter. As the industry develops and matures, new opportunities will emerge and new regions, including northwest Pennsylvania, will begin to see more direct benefits from shale gas development. New shale gas resources will help to stimulate midstream and downstream sectors in transportation, logistics, and manufacturing that have fewer direct impacts to gas drilling and exploration. In addition, cheaper energy will have other indirect benefits by reducing energy costs for homeowners and industrial users.

Within the Marcellus region, reduced natural gas prices have caused a shift of focus from "dry gas" fields (natural gas only), located in the eastern Marcellus Shale, to "wet gas" fields (which produce both oil and natural gas) in the western part of the region (Seydor et al. 2012). Because northwest Pennsylvania sits above both the Marcellus and Utica Shales, it is well positioned to take advantage of the ongoing activities in both formations. Moreover, the region is located near Shell Oil's ethane cracker facility now under construction in Potter Township, Beaver County (see chapter 7). When completed, this facility will separate ethylene from the ethane extracted from oil and gas wells. Ethylene is a foundational component for a wide variety of plastics products, so this development is expected to open many new manufacturing opportunities (IHS Markit 2017). Similarly, new pipelines and other infrastructure, such as the Mariner East Pipeline linking shale gas supplies in Western Pennsylvania to refineries in Eastern Pennsylvania, are expected to further stimulate demand in the region (Econsult Solutions 2018).

These trends will generate local opportunities to take advantage of the oil and gas activity underway in the region. In order to do this, however, firms must understand where those opportunities are located and where they fit within the broader oil and gas industry supply chain. The next section will take a more in-depth look at the structure of the industry and will map its supply chain.

Industry Structure and Value Chain Analysis for Northwest Pennsylvania

The basic supply chain for the Marcellus Shale development is based around ten primary areas (Seydor et al. 2012):

- Exploration
- Leasing, acquisition, and permitting
- Site construction
- Drilling
- Hydraulic fracturing
- Extraction and production
- Transportation and processing
- Storage
- Distribution
- Marketing

For the purposes of this analysis, we consolidated the Marcellus Shale supply chain into four primary components: site preparation, drilling and extraction, midstream, and downstream. Site preparation focuses on the exploration for potential drilling sites, and the subsequent acquisition and leasing of that drilling and extraction phase involves the actual drilling of the well, the hydraulic fracturing required to free the oil and gas from the shale, and then the subsequent extraction and production of oil and gas. The midstream component of the supply chain centers on the transportation, storage, and distribution of oil and gas. Downstream markets include manufacturing sectors, especially in plastics, where natural gas inputs are an important driver of production costs.

Proximity to gas deposits and drilling locations can bring direct economic benefits. However, an increase in oil and gas drilling can also create increased demand for a variety of goods and services. The resulting direct and indirect impacts are uneven across geographies and industry sectors. We use this framework as a starting point for our analysis and then attempt to connect these activities to specific industries. In doing so, we can gauge the nature and extent of these activities in northwest Pennsylvania and how they compare to the rest of Pennsylvania and the United States. We are specifically interested in the period from 2006 to 2016, which included the rapid expansion of gas production from the Marcellus Shale in Pennsylvania.

To identify the industries involved in those broad supply chain areas, we use the North American Industry Classification (NAICS) System (Executive Office of the President 2017). Much information was developed from interviews and literature reviews, with further analysis focused on the interindustry linkages of the core industries driving the oil and gas industry. Our analysis used an approach developed by Feser (2003). It provides a benchmark model of suppliers and markets for each industry, and traces interindustry connections. This approach designates forward linkages (sales made by the core industry, or market opportunities) or backward linkages (purchases made by the core industry, or supplier opportunities).

Data used in this analysis were developed by Economic Modeling Specialists International (2017). EMSI data track labor market information using both real-time sources and traditional data sets from federal agencies such as the US Department of Labor and the US Department of Education's National Center for Education Statistics. The EMSI data set is available by subscription and is widely used by economic development and workforce development organizations. It offers several benefits. It is often more timely than official government data, and the data set also limits the impacts of data suppression in smaller rural regions, which was an especially important attribute for tracking economic data in northwest Pennsylvania.

The diagrams presented below show the value-chain relationships found in the three primary components of the Marcellus Shale development, site preparation, drilling and extraction, and midstream. By considering the employment trends for each of the industries found within the value chain, we can begin to identify areas of potential growth and regional advantage.

Site Preparation

Figure 10.3 depicts the key industries associated with the site preparation phase of the Marcellus Shale development. Identifying potential sites often involves

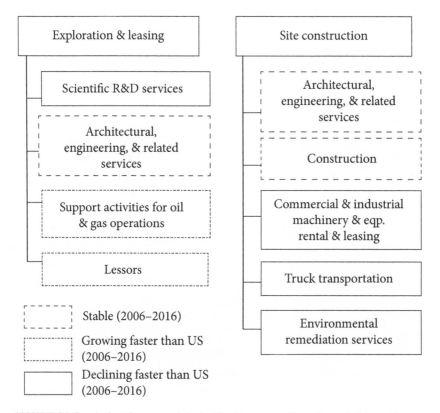

FIGURE 10.3. Industries associated with site preparation. Compiled by authors.

companies that fall within the research and development services in the physical sciences industry. This is a relatively small industry within the region, and these firms are generally headquartered elsewhere. Before drilling can begin, the drilling companies must obtain permission of the landowner. The leasing of land and mineral rights has created some business for local law firms, but this activity generates only a few local jobs. The significant amount of permitting, especially around environmental regulations and other compliance issues, has also generated demand for land services, including surveying, mapping, and technical support for environmental impact assessments.

Once the leasing agreements have been completed, the drilling company can begin to prepare the actual drilling site. This involves a significant amount of both engineering services and construction. In northwest Pennsylvania, employment in the engineering services industry has held steady over the past ten years—even as statewide engineering service employment has grown. However, it is important to note that this work is often undertaken by companies based outside the state and the region and therefore this activity is not completely captured by these employment data.

Preparing the drilling site also leans heavily on the construction industry. Beyond actually building the drilling pad, there is a need to either build or upgrade roads in order to get to the drilling site. Similarly, construction companies are needed to dig mud ponds to hold the water and sediment from the drilling process and install storage tanks for some of the chemicals and wastewater. In addition, sites require other features like fencing and temporary office buildings. Previous analysis of the period 2008 to 2013 highlighted several site preparation related sectors, such as truck transportation, that had enjoyed robust growth. When the period of analysis was extended to ten years (2006 to 2016), few sectors in northwest Pennsylvania enjoyed growth rates faster than those found in the state or the United States as a whole. Over this decade, only support activities and lessor sectors grew faster in northwest Pennsylvania. As more extensive drilling efforts resume, we might expect resumed growth in fields like truck transportation.

Truck transportation is vital to all parts of the drilling process. It is needed to move rock and dirt during the site preparation phase, water and chemicals during the drilling phase, and then transporting the actual oil and gas to distribution and processing facilities. Again, much as in the construction industry, trends in the truck transportation industry are not driven entirely by oil and gas. Nationwide, the truck transportation industry held steady between 2006 and 2016, declining only 0.1 percent. By contrast, declines in oil and gas production and manufacturing partially contributed to the region's truck transportation industry losing six hundred net jobs between 2006 and 2016, a loss of 1.4 percent annually over that decade.

Drilling and Extraction

Figure 10.4 shows some of the industries associated with the drilling and extraction part of the shale development process. Prominent industries in this segment of the process include drilling oil and gas wells, mining and oil and gas field equipment manufacturing, and construction. Drilling is a labor-intensive process that can help spur important ripple effects in the local economy. Additional workers create opportunities for hospitality, lodging, retail, and other services. During Pennsylvania's fracking boom, the drilling oil and gas wells sector was among the region's fastest growing industries. Lower oil and gas prices during the middle of this decade have meant lower employment at the national, state, and

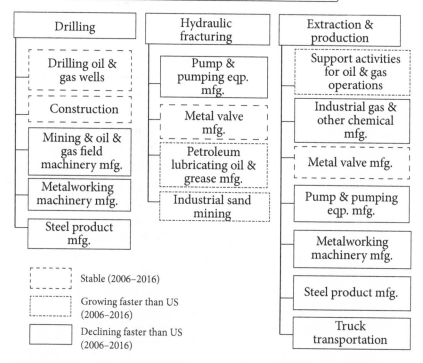

FIGURE 10.4. Industries associated with drilling and extraction. Compiled by authors.

local levels. Even so, with drilling companies mostly headquartered outside the region, local employment in the drilling sector remains small, with just over seventy jobs in 2016, and much lower than other industries within the supply chain.

Within the drilling and extraction supply chain—the core of the industry—companies fit within different tiers of suppliers. These companies include large oil and gas exploration companies, such as Anadarko Petroleum Corporation or Range Resources, which are typically headquartered outside Pennsylvania and rely on workers from other states to support drilling and other activities. Tier 1 suppliers provide products or services directly to these companies. Firms within the "support services to oil and gas operations" industry are often Tier 1 suppliers and may also be large firms. Halliburton, for instance, is active in the drilling and fracking process and would be considered a Tier 1 supplier. Tier 2 suppliers are those companies that either provide direct products or services to Tier 1 suppliers or indirectly sell products and services to the drilling and exploration companies. A company like Schramm (based in West Chester, PA), which produces mobile drilling rigs for oil and gas development, would be considered a Tier 2 supplier. Tier 3 suppliers are those firms that sell directly to Tier 2 companies but are more likely to be indirect suppliers to a wide range of oil and gas companies as well as companies in other industries in support activities such as transportation and logistics.

Firms operating in shale gas related supply chains operate across Pennsylvania, with large concentrations of companies in Pittsburgh, and in locations such as Williamsport and Canonsburg that provide proximity to drilling activities in the state's northern tier and southwest, respectively. A diverse mix of firms and specializations can be found in these locations. Within northwest Pennsylvania, local employers typically specialize in Tier 2 or Tier 3 activities in areas such as leasing, engineering, and technical services that go into acquiring leases and planning drilling locations. Construction and site preparation activities also account for a sizable share of local industry employment.

The manufacturing of machinery used in mines and oil and gas fields was an area of greater strength for the region. In 2013, the region had almost 1,300 jobs in this industry, but this total employment level has now been slashed to 743 people thanks to downturns in both gas and coal activities in the state. Even with cutbacks, this industry is also highly concentrated in the region, but has not grown as fast as the national industry average. However, this sector could offer great potential for northwest Pennsylvania's smaller manufacturers as domestic oil and gas activities continue to grow.

Both the fracking and extraction processes generate demand for other manufactured goods. For instance, the drilling and extraction phases make extensive use of pumps and pumping equipment, valves and pipes, and metalworking machinery (which includes cutting and machining tools). Two of the industries involved in manufacturing these products—metal valve and pipe manufacturing and metal working machinery manufacturing—are highly concentrated in the region. However, their recent growth trajectories have differed over the past decade. The region's valve and pipe manufacturing industry—which employs more than 1,100 people in the region—grew by less than 1 percent annually between 2006 and 2016, even though the industry nationwide lost net employment. By contrast, the region's metalworking machinery manufacturing industry—which accounted for 2,500 jobs regionally in 2016—lost employment at a rate more than twice the national average, largely due to downsizing at local facilities such as Erie's GE Transportation facility.

Because drilling and extraction use steel as a critical input, the steel industry benefits from growth in the oil and gas industry. The steel industry has lost employment in northwest Pennsylvania over the past five years; yet it still employs more than a thousand people and remains roughly nine times more concentrated in the region than in the US economy as a whole. The demand generated by the oil and gas industry for pumps, pipes, tanks, metalworking machinery, and many other steel-dependent products has likely benefited the steel industry even as it has suffered net job loss.

The drilling and extraction phase of the oil and gas supply chain also creates demand for other industries. For instance, companies in the industrial gas and other chemical manufacturing industry are responsible for some of the fluids in the fracking process. The processes also rely upon the petroleum lubricating oil and grease manufacturing industry. In both instances, employment within northwest Pennsylvania is minimal. Combined, both industries account for fewer than two hundred jobs. The drilling and extraction process also requires ample supplies of "frac sands." Industrial sand is another key input in the fracking process, and the region has no industrial sand mining employment. Much of the frac sands come from Wisconsin, where the sand is the right material (quartz), shape, and size and the sand deposits are located near bulk transportation corridors such as barge and rail (Deller and Schreiber 2012).

Midstream Activities

Midstream activities are those that involve the transportation, storage, and distribution of oil and gas products. These activities have typically supported the exploration and production companies, but the recent shale-gas development has also created significant demand for midstream companies. For example, analysts estimate that 35 percent of the gas drawn from the Bakken shale play needs to be burned off because of insufficient infrastructure to store or transport it (Deloitte Center for Energy Solutions 2013). As the domestic oil and gas industry develops, midstream companies can capture new markets in midstream sectors like storage and distribution.

Across the industry, few firms have a major presence in both production and midstream activities. Key midstream firms include national players such as Mark-West Energy, Williams Midstream Services, and Energy Transfer Partners. The midstream sector involves distinct local, regional, and national activities. The local segment involves the oil and gas gathering and processing systems required in the actual oil and gas fields, and, as a result, demand for this infrastructure is greatest when new shale wells are being drilled and production levels are high. The regional and national segments are the pipeline networks that transport oil and gas products around the country. They are the equivalent to the highways of the interstate highway system.

Figure 10.5 shows the industries associated with midstream activities. Truck transportation remains as a primary method for moving oil and gas, and it has

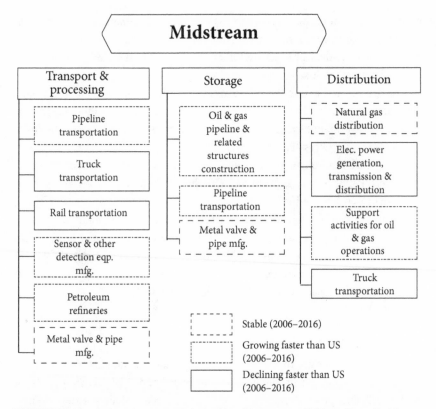

FIGURE 10.5. Industries associated with midstream activities. Compiled by authors.

experienced growth over the past five years. Pipeline transportation is another midstream opportunity that is on the cusp of a major transformation. It is a small industry within northwest Pennsylvania (153 jobs), but, between 2006 and 2016, its regional growth rate (6.3 percent) far outpaced that of the nation (2.4 percent). The development of new storage and transmission infrastructure, such as the Mariner East pipeline, will also boost local jobs in coming years. As it has grown, the industry has also become increasingly concentrated. While still small in terms of total jobs, the industry is 60 percent more concentrated in the region than in the United States as a whole. Employment in construction specifically related to oil and gas pipelines has also grown rapidly over the past decade (7.4 percent annually), but still accounts for a small number of total jobs.

In addition to pipelines, rail provides shipping for much of the region's oil and gas. Due to this increased demand, CSX is making major investments in the region's rail infrastructure. The rail transportation industry currently accounts for about 471 jobs in the region, down from 544 jobs in 2006. This stands in contrast to both Pennsylvania and the nation, which have experienced modest growth during the same period. Nonetheless, freight rail is rightly viewed as a major future midstream opportunity. Rail terminals can be built relatively quickly compared to similar pipeline facilities. Moreover, rail can service every US refinery and offer flexibility to the oil and gas industry as it shifts production between different shale plays. Within northwest Pennsylvania, this opportunity also aligns well with the region's historical strengths as a center for both rail transportation and rail-related manufacturing opportunities.

Several other industries also merit mention. Natural gas distribution has experienced net employment growth both nationwide and regionally. This sector, accounting for about 431 regional jobs, is still twice as concentrated in the region when compared to US averages. Further downstream, power plants that produce, transmit, and distribute electricity also lost employment between 2006 and 2016, at a slightly faster rate than national averages. Midstream activities also generate their own spin-off opportunities and spur new demand for jobs related to surveying, civil and environmental engineering, and environmental assessments. They can also create jobs in construction and other trades like welding that are essential to the construction and maintenance of this equipment and infrastructure. They also spur demand for materials like steel and components like valves, pumps, and sensors. As noted above, the region has some strength in valve manufacturing, but pump and pumping equipment manufacturing is small and declining. There is currently no regional employment in the sensors and other detection equipment manufacturing industry. All of these sectors could grow as local shale gas development increases.

Downstream Opportunities

While these direct and midstream opportunities are generating jobs and local economic activity, more sustained and long-term impacts are likely to be found in manufacturing activities that benefit from proximity to shale gas resources, and more generally from the emergence of cheaper energy inputs into a range of manufacturing industries. Northwest Pennsylvania is well positioned to compete and prosper on both accounts as it has easy access to emerging shale gas resources and is also home to a large base of manufacturing industries, many of which can and will benefit from lower energy costs.

The chemical industry is among those expected to enjoy the greatest benefits from the shale gas revolution. In fact, in February 2014, new industry investment triggered by the shale gas revolution exceeded $100 billion, with more than 637,000 potential new jobs tied to these investments (American Chemistry Council 2014). These effects are expected to ripple across the chemical industry value chain and generate new market opportunities in dozens of related sectors such as plastics, coatings, tires, textiles, and paints.

In Pennsylvania, much attention is focused on Shell Chemical's ethane cracker facility in Beaver County as discussed in chapter 7. This multi-billion-dollar facility, the first of its kind in the region, is expected to generate many new business opportunities, along with massive job impacts related to plant construction. According to research from IHS Markit (2017), the facility is ultimately expected to produce 1.5 million tons of polyethylene per year.

Numerous industries, from packaging to clothing to adhesives, operate downstream in the ethylene supply chain. Through the steam cracking process, ethylene is separated from the ethane extracted from oil and gas wells. Ethylene is then turned into intermediate products like PVC and polystyrene which are then used in a wide array of different products. Among the industries that might benefit from being proximate to this facility include chemicals, plastics, and other petrochemical by-products.

This project will have massive impacts and is expected to create significant opportunities for Western Pennsylvania and eastern Ohio. By locating in the Marcellus region, the Shell plant will allow manufacturers to better manage costs within their respective supply chains. For instance, there are several polymer and converter companies in the region that turn polymers into finished products or components for other goods, such as automobiles. The Shell facility will allow them to reduce the time and costs involved with procuring a key input.

Even with this massive production expansion, industry experts believe that prospects for further production expansion are good. A recent IHS Markit (2017) analysis prepared for the Team Pennsylvania Foundation suggested that the region could accommodate as many as four additional facilities of similar scale and scope to the Royal Dutch Shell facility. This projection is based on industry estimates of growing demand, and the region's location advantage. It is located within seven hundred miles of two-thirds of the current US and Canadian industrial demand for polyethylene and polypropylene, a specialized resin used in multiple industries.

While the cracker facility is under construction, companies and communities need to begin exploring strategies for taking advantage of the opportunities that may become available through this development. When this facility comes online, it will be producing for industries and companies well beyond Western Pennsylvania, but there should be opportunities for local firms.

Oil and gas refining is an industry that can fall in either the midstream or downstream segment of the oil and gas supply chain. Within the region, petroleum refinery employment grew by 1.4 percent annually between 2006 and 2016. This employment, however, remains relatively small with just under four hundred jobs in the region. Given the large capital investments required with building these kinds of facilities, it is more likely that the region will export its oil and gas to places with large petrochemical complexes like Houston and the Gulf Coast. The expansion of the pipeline infrastructure will further facilitate this process. As noted earlier, however, the real benefit for manufacturers will come from the lower energy costs, particularly in energy-intensive industries. Lower costs can help reduce operating costs and allow existing regional manufacturers to expand and invest in their current facilities.

While the chemical industry is widely expected to grow thanks to new shale gas resources, it is not alone. A recent Peterson Institute analysis assessed the projected impacts of new oil and gas investment on US manufacturing sectors (Houser and Mohan 2014, 82–94). It found the greatest positive employment and output impacts in the following industries:

- Cutting tool and machinery tool accessory manufacturing
- Mining and oil and gas field machinery manufacturing
- Steel product manufacturing
- Air and gas compressor manufacturing
- Fabricated pipe and pipe fitting manufacturing
- Pump and pumping equipment manufacturing

All of these sectors are projected to see major demand increases, along with large infusions of new capital (Houser and Mohan 2014). Other industries, such as fertilizer and chemical sectors, will see most benefits from a reduction in energy costs as opposed to new customer demands.

Supply Chain Issues and Opportunities

The oil and gas industries and related sectors provide real opportunities for local sourcing. Many firms involved in these supply chains are interested in local sourcing opportunities to shorten delivery times and increase responsiveness. However, firms do not just stumble into these opportunities, and they likely need to do some self-assessment to ensure that they can meet the industry's exacting needs. Much like the aviation industry, firms in the oil and gas industry have high standards of their suppliers, and they will quickly replace underperforming vendors. To compete and win in these new markets, potential subcontractors must consider several key issues.

First, quality is paramount. The activities associated with oil and gas exploration, extraction, and distribution require a great deal of precision. Equipment used in exploration and drilling faces rigorous conditions, and the costs of equipment failure—in terms of both time and money—can be high. This is especially the case for activities related to drilling and exploration that often take place overseas and in out-of-the-way places where replacement parts may be days away rather than hours.

How can firms fully embrace high quality? Firms looking to supply manufactured products or services to the oil and gas industry can demonstrate the quality of their products by investing in relevant industry certifications. In addition to prominent industry certification programs like ISO 9001, the American Petroleum Institute has developed additional certifications that reflect the industry's exacting quality standards. These certifications include API Spec Q1 for manufactured products and API Spec Q2 for service providers, along with a host of other specialized certifications such as those that focus on the traceability of steel and other factors.

While certifications are one important way in which oil and gas companies can identify quality vendors, other tools also exist. Some OEMs and other leading companies also view participation in other leading supply chains as another indicator of quality. For instance, one manufacturer of drilling rigs noted that if a vendor was selling to leading companies, such as Boeing or Caterpillar, then they would likely be able to meet their quality standards. Participating in these multiple value chains can benefit companies because it helps them diversify their customer base and buffer them against the ebbs and flows of any one given industry.

Second, firms must also have strong safety, environmental, and health standards, along with strong record keeping to document and demonstrate these standards. These quality and safety requirements can be a significant impediment for small firms looking to take advantage of oil and gas opportunities. The challenge is not so much that they cannot meet the desired levels of quality and safety. Rather, they often lack the resources and staff capacity to comply with all the rules and documentation used by OEMs and they cannot afford the high costs for regulatory compliance. Market success requires investment in these critical compliance resources and capacity.

Third, firms must also remain flexible and maniacally focused on committing to meeting deadlines. Vendors must be available 24/7 to meet client needs. If that means delivering a part or a service at 2:00 a.m., then that is what they must do. Beyond this level of short-term responsiveness, firms must be able to deliver on actual deadlines as well as their forecast planning. Firms working within the oil and gas industry may also want to consider pursuing "Master Service Agreements" with major oil and gas companies as a way to better position themselves for this work. Basically, these agreements allow vendors to be recognized as pre-approved vendors. In doing so, they are more likely to be selected to do work when production increases or expands.

Before entering into oil and gas supply chains, firms must also bolster their financial position. Financial solvency is an important consideration when selecting vendors because payments may not turn around as quickly as in other industries. For instance, suppliers may get paid in ninety days rather than thirty, perhaps creating cash flow challenges for smaller firms. In order to help them assess the financial solvency of potential vendors, many large oil and gas companies use online tools, like ISNetworld, to find potential suppliers and vendors. For smaller companies, supplying company information to these databases improves their chances of connecting to larger companies and securing potential work.

The industry's emphasis on responsiveness, quality, and safety all create opportunities for domestic producers. Many of these demands cannot be met by more inexpensive overseas producers, often creating a distinct preference for domestic producers. Many of the largest oil and gas companies are based in Texas and Oklahoma and have ongoing relationships with existing suppliers. Over time, more Pennsylvania-based companies have been able to capture these markets by demonstrating these key traits, and their proximity to the Marcellus Shale development has helped improve their responsiveness and reduce shipping costs. Moreover, local labor—rather than skilled laborers brought in from the Texas oilfields—is significantly less expensive. Smaller firms should consider outreach to supplier companies before reaching directly to the large national market leaders. They should first understand their place in the oil and gas supply chain and use that as a key input in crafting their strategy to enter the supply chain.

Recommendations

The potential for a manufacturing renaissance in northwest Pennsylvania is strong. Even as manufacturing firms and jobs have exited from other parts of the United States, northwest Pennsylvania has retained a robust local manufacturing base operating in a diverse mix of sectors.

The emergence of cheaper shale gas is a potential game changer for manufacturing firms across the United States, especially in Rust Belt states still confronting challenges with deindustrialization as traditional manufacturing declined. Northwest Pennsylvania is well-positioned to benefit from these new resources— thanks to its location near the Utica and Marcellus shale plays, and its past success in a diverse set of manufacturing industries.

Trends in leading manufacturing sectors are generating exciting potential opportunities. However, at this point, it is best to speak of "potential" opportunities as opposed to "sure things." Many of the critical fundamentals are in place, but local leaders, corporate executives, and local manufacturers must act aggressively to build on this potential. Below, we offer recommendations that can help build a stronger base for manufacturing across northwest Pennsylvania.

Support Development of Essential Infrastructure

Rail transportation and shale gas developments can advance together. Much of the current growing demand for freight rail is driven by growing use of rail as a means to transport shale oil and gas. Continued investment in regional rail infrastructure can help support further industry growth by increasing demand for new rail cars and equipment, expanding broader demand for freight rail, and reducing transportation costs for local manufacturers. This effort, along with continued investments in warehousing and distribution infrastructure, can help further solidify northwest Pennsylvania's position as an important regional logistics hub.

Tap Critical Federal and State Funding Streams

As the movement to promote the reshoring of manufacturing continues, we can expect an increase in new investments from federal, state, and local governments. Regional leaders must position themselves to compete for and win existing and new investment opportunities. Key regional partners already have a good track record of securing outside investments, and continued focus on this important work is needed. In addition to supporting regional economic and workforce development programs, regional leaders should also aggressively tap into outside funds that can help spur further development in key manufacturing sectors such as rail and transportation and in sectors related to shale gas development.

Build on Key Cluster Strengths

In recent years, several regions and states have begun more concerted efforts to focus public attention and resources on these important manufacturing-related sectors. For example, the Marcellus Shale Coalition (Pennsylvania), the Northwest Pennsylvania Oil and Gas Hub, and the Ohio Oil and Gas Association have emerged as strong advocates in the shale gas sector, advocating for the development of new infrastructure and training programs to support industry expansion.

In an effort to further spur growth in key manufacturing sectors, regional and local economic development agencies should support the development of regional networks to support and advocate for these industries. These initiatives can take the form of new networks focused on specific manufacturing sectors or could be connected to existing networks for small- and medium-sized manufacturers. Programs of this type are supported by state initiatives, such as the Industrial Resource Center Network, a local affiliate of the National Institute of Standards and Technology Manufacturing Extension Partnership (NIST-MEP) program. This network helps small manufacturers with issues such as retooling, new market development, and certifications.

These new networks should focus on several important functions. First, they should help firms identify and capture new market opportunities at home and abroad. Second, they should build stronger supply chain connections. This task typically involves a mix of effective publicity and communication along with enhanced transparency and connections between firms within the supply chain. Regular network events to connect suppliers and learn about industry trends are an important first step in this work. These efforts should also embrace the use of new online supply chain connection tools. For example, the Manufacturer's Marketplace (National Association of Manufacturers 2021) and the Connectory (San Diego East County Economic Development Council 2021) are nationwide web-based networks that contain profiles of companies in a host of industries and levels of the supply chain. In addition to its national database, the Connectory also includes specialized regional networks such as defense and aerospace suppliers located in San Diego or in the Pacific Northwest. Finally, networks can help firms and industries deal with pressing workforce development challenges. Finding a qualified workforce has been a major challenge across Pennsylvania, particularly in the fast-growing sectors related to shale gas development.

These regional efforts should be supplemented with support programs targeted to the local level. Communities in the first wave of Marcellus Shale development, such as Washington and Lycoming Counties, have developed excellent local initiatives to support industry development. These types of initiatives are now being developed in northwest Pennsylvania. For example, the Northwest Oil and Gas Hub Taskforce sponsored a wide range of local and regional events. The community of Titusville is also presently completing a strategic plan to help guide its own local development of shale gas resources.

Enhance Overall Business Support Service Capacity

While many local manufacturers face challenges unique to operating in shale gas related industries, many of their growth challenges result from other less-industry-specific factors. Like entrepreneurs in other sectors, manufacturers need access to sophisticated consulting and business support tools.

Thus, these regional networks need to do a better job of linking local manufacturers to high-quality business development support, technical support from university researchers, and outside investment options. In some cases, firms within these manufacturing sectors should be connected to existing and new NIST MEP programs such as ExporTech or supply chain optimization. Connections to financing are also crucial. Firms entering new markets will need outside investment to purchase new equipment, provide necessary training, and obtain needed certifications and the like. Without new infusions of working capital, their ability to retool could be hampered.

Embrace Excellence across Supply Chains

These manufacturing sectors offer real opportunities for local sourcing, but only if local suppliers adhere to high standards of responsiveness, quality, safety, and fiscal stability. Being local is not enough. Vendors must embrace quality and excellence if they hope to develop new and profitable supply chain connections.

Small business support to help manufacturers achieve industry recognized certifications could bring real benefits to many local firms. For small firms, engaging with larger OEMs typically requires compliance with industry recognized standards. Those standards vary widely from industry to industry. In general, companies use certifications as a third-party validation of the work processes and likely product outcomes. For example, the well-known International Standards Organization (ISO) through its ISO 9000 standards certifies a firm's quality management procedures and practices. For certain industries, ISO certification is sufficient, but, for others, the certifications or regulatory requirements can be quite specialized. For instance, an AS9100 provides a basic quality standard for the aerospace industry.

Obtaining these certifications and registrations can be cost prohibitive and create a serious barrier to small companies seeking to capture new supply chain opportunities. Assistance and funding support to pursue certifications may help local firms compete in these new markets. This type of assistance has been usefully provided in a number of states, such as North Carolina and Arizona.

References

American Chemistry Council. 2013. "Shale Gas, Competitiveness, and New U.S. Chemical Industry Investment: An Analysis Based on Announced Projects." https://www.americanchemistry.com/First-Shale-Study/.

American Chemistry Council. 2014. "U.S. Chemical Investment Linked to Shale Gas Reaches $100 Billion." https://www.americanchemistry.com/Media/Press ReleasesTranscripts/ACC-news-releases/US-Chemical-Investment-Linked-to-Shale-Gas-Reaches-100-Billion.html.

Cruz, Jennifer, Peter Smith, and Sara Stanley. 2014. "The Marcellus Shale Gas Boom in Pennsylvania: Employment and Wage Trends." Monthly Labor Review. US Bureau of Labor Statistics, February. https://doi.org/10.21916/mlr.2014.7.

Deller, S., and A. Schreiber. 2012. "Frac Sand Mining and Community Economic Development." University of Wisconsin–Madison. Department of Agricultural and Applied Economics. Staff Paper no. 565. https://econpapers.repec.org/paper/eclwisagr/565.htm.

Deloitte Center for Energy Solutions. 2013. The Rise of the Midstream: Shale Reinvigorates Midstream Growth. Houston and Washington, D.C.: Deloitte Center for Energy Solutions. https://www2.deloitte.com/be/en/pages/energy-and-resources/articles/the-rise-of-the-midstream.html.

Economic Modeling Specialists International. 2017. http://www.economicmodeling.com/data/.

Econsult Solutions. 2018. "The Potential Economic Impacts of the Mariner East Pipelines." The Revolution Project Fractionation Facility, and Associated Improvements at the Marcus Hook Industrial Project. January 8. https://econsultsolutions.com/the-potential-economic-impacts-of-the-mariner-east-pipelines.

Executive Office of the President of the United States. Office of Management and Budget. 2017. North American Industry Classification System. https://www.census.gov/eos/www/naics/2017NAICS/2017_NAICS_Manual.pdf.

Feser, E. J. 2003. "What Regions Do Rather Than Make: A Proposed Set of Knowledge-Based Occupation Clusters." Urban Studies 40 (10): 1937–58.

Genedge Alliance. 2012. "U.S. Supply Chain Competitiveness: How Agility, Planning, Collaborative Product Development and Supplier Alignment/Sourcing Execution Are the Keys to Profitable Growth." Report prepared for NIST-MEP Supply Chain Development Initiative. Martinsburg, VA: Genedge. https://silo.tips/download/prepared-by-genedge-alliance-nist-mep-supply-chain-development-initiative-cooper.

Gray, Wayne, Joshua Linn, and Richard Morgenstern. 2018. "The Impacts of Lower Natural Gas Prices on Jobs in the U.S. Manufacturing Sector." Washington, DC: Resources for the Future. https://media.rff.org/documents/RFF20Rpt20Gas 20Prices20Jobs.pdf.

Houser, T., and S. Mohan. 2014. Fueling Up: The Economic Implications of America's Oil and Gas Boom. Washington, DC: Peterson Institute for International Economics.

IHS Markit. 2017. "Prospects to Enhance Pennsylvania's Opportunities in Petrochemical Manufacturing." Executive Summary. https://teampa.com/wp-content/uploads/2017/03/Prospects_to_Enhance_PAs_Opportunities_in_Petrochemi cal_Mfng_Report_21March2017.pdf.

McKinsey Global Institute. 2012. "Manufacturing the Future: The Next Era of Global Growth and Innovation." https://www.mckinsey.com/business-functions/operations/our-insights/the-future-of-manufacturing#.

Nager, Adams. 2017. "Trade vs. Productivity: What Caused U.S. Manufacturing's Decline and How to Revive It." Information Technology and Innovation Foundation. February. https://itif.org/publications/2017/02/13/trade-vs-productivity-what-caused-us-manufacturings-decline-and-how-revive.

National Association of Manufacturers. 2021. Manufacturers Marketplace. Washington, DC: National Association of Manufacturers. https://manufacturersmarketplace.us/.

San Diego East County Economic Development Council. 2021. Connectory. El Cajon, CA: San Diego East County Economic Development Council. https://www.connectory.com/.

Seydor, S., E. Clements, S. Pantelemonitis, V. Deshpande, et al. 2012. "Understanding the Marcellus Shale Supply Chain." University of Pittsburgh. Joseph M. Katz Graduate School of Business. https://marcelluscoalition.org/wp-content/uploads/2012/08/Pitt_Understanding_the_Marcellus_Shale_Supply_Chain_May_2012.pdf.

Sirkin, Harold, Justin Rose, and Michael Zinser. 2012. "The U.S. Manufacturing Renaissance." Boston Consulting Group Report. November.

StateImpact.org. "What's an Ethane Cracker?" https://stateimpact.npr.org/pennsylvania/tag/ethane-cracker.

THE BOOM, THE BUST, AND THE COST OF THE CLEANUP

Nicholas G. McClure, Ion G. Simonides,
and Jeremy G. Weber

Pennsylvania has the longest history of oil and gas extraction in the world, with almost 160 years of drilling and production. Over time, governance of extraction has matured but continues to evolve, especially regarding bonding and reclamation of wells that have aged beyond their economically useful life. The long history of policy development provides a panoramic perspective useful for the governance challenges of ensuring responsible oil and gas development in the shale era.

We make three observations from the Pennsylvania experience. First, sensible policies can be slow to emerge, especially those governing well reclamation—the removal of well equipment, the filling of well holes, and the restoration of well sites. Statutes concerning well reclamation were enacted before 1900, but permitting requirements necessary for enforcing reclamation emerged a century after the first oil well was drilled. A second observation is that the mere presence of reclamation requirements and knowledge of well locations does not ensure proper reclamation. We estimate an abandonment rate of about 20 percent for wells drilled under a regulatory regime that required reclamation but did not ensure that firms had the finances to do so. The limitation of reclamation requirements is underscored by numerous times in the state's history where drilling boomed along with a genesis of new firms only to be followed by times where oil and gas prices did not cover the cost of production, causing the demise of many firms. And as the dead don't pay taxes, defunct firms don't reclaim wells even if required by law.

Last, conventional well bonding requirements, which oblige well operators to set aside money that is only released after proper well reclamation, have been far less than reclamation costs (roughly half of the costs by our estimates). The same is now true of shale well bonding requirements. A single shale well bond is $10,000, but our analysis of reclamation costs for more than 1,200 wells suggests that reclaiming a typical shale well could plausibly cost more than $90,000. Although bonding requirements are not the only means for encouraging and paying for proper reclamation, other methods, such as setting aside revenue raised through the state's impact fee on unconventional wells, are not currently in effect.

Abandoned Wells: Definitions, Estimates, and Risks

Current Pennsylvania regulations hold the firm that operates a well responsible for reclaiming it, including plugging it and restoring the well site to predrilling conditions. We use the term "abandoned well" to mean any well where the firm ceases to maintain the well with the intent to produce from it and has not reclaimed it according to state law. More technically, the Pennsylvania Department of Environmental Protection (DEP) defines an abandoned well as any of the following: a well that has not been used to produce, extract, or inject any gas, petroleum, or other liquid within the preceding twelve months; a well for which equipment necessary for production, extraction, or injection has been removed; or a well considered dry and not equipped for production within sixty days after drilling, redrilling, or deepening (58 Pa. Cons. Stat. §2301 et. seq.). If a firm fails to reclaim a well, then it is considered abandoned. Based on state records and the analysis of historical documents, estimates of the number of abandoned wells in Pennsylvania range from 300,000 to 500,000 (Kang et al. 2014; Dilmore et al. 2015).

Abandoned wells can provide pathways for water and air contamination by leaking brine, oil, or methane, thereby posing risks to people, local ecosystems, and the global environment. Several cases illustrate the point. As Royal Dutch Shell subsidiary East Resources was drilling a shale gas well in June of 2012, methane began to bubble out of nearby streams and the water well of a neighboring cabin overflowed, flooding the cabin. The unintended migration of methane seemingly caused by Shell's drilling culminated in a thirty-foot geyser of water and natural gas erupting out of a nearby abandoned well for over a week. The abandoned well had been drilled as a gas well in 1932, and while Shell knew of the well's proximity to their operation, they assumed that it had been properly plugged. Shell asked local residents to evacuate their homes while they worked with well control specialists, a fire department, and state regulators to get the leak under control (Detrow 2012a).

A similar, but more violent, incident occurred in 2011 in Bradford Township, McKean County. Marcellus Shale gas drilling was again considered the likely cause of methane migration through century-old abandoned wells resulting in two nonfatal home explosions (Boose 2011). Aside from incidents related to recent shale gas development, improperly plugged wells can leak methane into nearby homes over time. A 2014 *Pittsburgh Post-Gazette* article described the accumulation of dangerously high levels of methane under homes in Versailles Borough, Allegheny County, due to improperly abandoned natural gas wells dating from the 1920s (Litvak 2014). Much of the evidence regarding local problems associated with abandoned oil and gas wells is anecdotal or focused on particular places or cases.

More systematic studies of abandoned wells have focused on methane emissions and their implications for global warming given that methane is a more potent greenhouse gas than carbon dioxide. Townsend-Small et al. (2016) sample 138 abandoned oil and gas wells across the United States and find that 6.5 percent had measurable methane emissions. Focusing on Pennsylvania, Kang et al. (2016) estimate that abandoned wells account for 5 to 8 percent of annual anthropogenic methane emissions in the state. In addition to trapping heat in the atmosphere, the oxidation of methane produces ozone locally, which can cause respiratory and cardiovascular health problems (Kang et al. 2014; Shen and Gao 2018).

Why Are There So Many Abandoned Wells in Pennsylvania?

Governance Pre-1984: The Beginning to Permitting and Reclamation Requirements

The first commercial oil well was drilled near Titusville, Pennsylvania, in 1859 by Edwin Drake. The discovery of oil quickly transformed Western Pennsylvania into the site of the world's first oil boom. The first decades of the industry were defined by massive influxes of capital, rapid technological development, and booms and busts. Understandably, the state had no regulations when the oil industry burst into existence in the 1860s, and firms abandoned wells whenever and however they wanted. The low upfront costs of drilling a well meant low barriers to entry into the industry, which attracted entrepreneurs to the region.

The industry quickly generated considerable wealth for some, but supply gluts and price crashes brought financial ruin for others. Two journalists, Andrew Cone and Walter R. Johns, in their history of the first decade of oil development in Pennsylvania describe how "oil companies fell, in the fall and winter of 1866 and 1867, one after another, like a row of bricks" (Cone and Johns 1870). Except in the case of a buyout by a competitor, a firm's responsibility for abandoned wells ended when a bankrupt firm dissolved.

The time of booms and busts and the time of small producers faded in the late 1880s as John D. Rockefeller started to monopolize the industry with his company Standard Oil. As the industry matured, so did governance. In 1881 the state passed a statute that required the plugging of abandoned oil wells and provided a penalty for not doing so (Ely 1933).[1]

Production declined throughout the early 1900s as the oil industry shifted westward to new reservoirs in states like Texas and California. As the number of abandoned wells mounted, the state passed another statute in 1921 requiring the plugging of all abandoned wells, not just oil wells. The statute was a major policy improvement, but an absence of well permitting requirements made it impossible to systematically enforce well reclamation.

The discovery of the Leidy natural gas field in Clinton and Potter Counties in 1950 led to a new boom in exploration and development that lasted through the decade. The H. E. Finnefrock well, drilled in 1951 by New York State Natural Gas (now Consolidated Natural Gas Corporation) bore the distinction of the largest natural gas well discovered east of the Mississippi (Waples 2014). According to one account of the discovery of the Leidy field, "Gas fever spread to young and old, and those who could spare it plunked down their money for stocks in a number of drilling companies with optimistic plans and hopes" (Skinner 1990). In the middle of the boom, the state enacted Act 225 of 1995, known as the Gas Operations Well-Drilling Petroleum and Coal Mining Act, to establish permitting and registration requirements for all new oil and gas wells. By giving the state the means to know the location and owner of any new well drilled, the act enabled enforcement of the reclamation statute passed in 1921.

The policy legacy helps explain the state's current data limitations regarding oil and gas wells. Because the state began permitting wells in 1957, it has limited information on wells drilled in earlier years. However, there is an ongoing effort at the Department of Conservation and Natural Resources (DCNR) to retroactively enter pre-1957 wells into their public online database, with approximately ten thousand wells entered so far. As Act 225 developed more detailed permitting regulations, the DEP and DCNR digital well records improve significantly for wells drilled after 1957.

TABLE 11.1 Summary of significant oil and gas statutes, Pennsylvania

YEAR ENACTED	ACT NAME OR NUMBER	SUMMARY OF SIGNIFICANT PROVISIONS
1881	Act 101	Required plugging of abandoned oil wells and provided penalty for operators who failed to do so.
1921	Act 322	Expanded plugging requirement and penalty to gas wells.
1955	Gas Operations, Well Drilling, Petroleum, and Coal Mining Act	Established permitting and registration requirements for all new oil and gas wells drilled after 1957.
1987	Oil and Gas Act	Increased permitting and registration requirements, required reporting of drilling and production by operators, and established bonding requirements for wells.
2012	Act 13	Revised oil and gas regulatory regime to provide for shale gas development and unconventional wells.

Governance Post-1984: Introducing Bonding Requirements

In response to higher prices, the late 1970s and early 1980s saw an expansion in natural gas production and modest increases in oil production. As in the past, changes in policy occurred with shifts in industry activity, and in 1984 the state passed Act 223 that repealed Act 225 of 1955 and increased the state's responsibility for permitting and regulating the industry. Act 223 required detailed reporting of drilling and production, the plugging of all wells, and instituted bonding requirements for wells drilled after April 17, 1985.

Bonding requires firms operating wells to set aside funds before drilling a well, which are forfeited to the state if the firm fails to comply with state standards. The financial incentive provided by bonding attempts to ensure that firms will properly perform drilling operations, resolve any water supply issues, and properly plug wells at the end of their economically useful life. Act 223 set the bond for a single well at $2,500. For multiple wells, a blanket bond covering an unlimited number of wells was set at $25,000 (Act 223 of 1984). The bond could take the form of a cash deposit, certificates of deposit, letters from financial institutions proving adequate lines of credit or collateralized assets equal to the amount of the bond. Operators receive interest accrued from collateral bonds and certificates of deposit. The bond is released by the state one year after the state receives a plugging certificate from the firm operating the well that documents proper reclamation (Act 223 of 1984).

The act also gave the Environmental Quality Board the authority to adjust bonding amounts every two years in line with inflation and changing reclamation costs (Oil and Gas Act 1984). The board never changed bonding requirements while the act was in effect despite having the authority to do so. The bond amount for a well drilled in 1985 was therefore the same as for a well drilled a quarter century later despite inflation and other economic forces causing the cost of plugging a well to rise. Evidence suggests the board should have exercised its authority to increase bonding requirements: we estimate that more than six thousand wells were likely abandoned during the twenty-seven-year period that the act was in effect.[2]

Because many abandoned wells have no identifiable owner (recall that the state did not require permits of wells for the first century of drilling), or the owner has subsequently dissolved, 1992 amendments to Act 223 of 1984 designated such abandoned wells as orphan wells. The act made orphan wells the responsibility of the Department of Environmental Protection, leaving landowners, leaseholders, and firms exempt from any responsibility if they received no economic benefit from the well after April 1979. The 1992 amendment also created the Orphan Well Plugging Program, which is financed by a fee on new well drilling permits and provides the department with approximately half a million dollars per year to reclaim orphaned wells (Pennsylvania Department of Environmental Protection 2000). As of 2019, the department had plugged 3,419 orphan wells (Pennsylvania Department of Environmental Protection 2019).

Passed in 2012, Act 13 (Title 58 of the Pennsylvania Consolidated Statutes §2301 et seq.) replaced Act 223 of 1984, updating bonding amounts and basing them on wellbore length and the number of producing wells. It also created distinct bonding requirements for shallow and deep wells, with deep wells generally corresponding to unconventional wells. The basis for the bond amounts is unclear but likely stems from recommendations from the Oil and Gas Technical Advisory Board, which includes an engineer, geologist, and technicians from the industry appointed by the governor (58 Pa. Cons. Stat. §3226). As in Act 223 of 1984, the Environmental Quality Board may adjust bond amounts every two years to reflect the projected costs to the commonwealth for reclaiming wells.

Bonding requirements under Act 13 split wells into two categories based on total wellbore length: less than 6,000 feet and greater than 6,000 feet, which roughly correspond to conventional and unconventional wells. Within the categories, bond amounts depend on the firm's number of wells. For example, a firm that operates fifty-five wells with wellbore lengths less than 6,000 feet pays $35,000 for the first fifty-one wells and an additional $4,000 per well for the remaining four wells, for a total amount of $51,000. If that same firm increases

TABLE 11.2 Bonding requirements for oil and gas wells under Act 13, Pennsylvania

TOTAL WELLBORE LENGTH	LESS THAN 6,000 FEET (MOSTLY CONVENTIONAL)			
Number of wells bonded	<51	51–150	151–250	>250
Flat fee	None	$35,000	$60,000	$100,000
Bond per additional well	$4,000	$4,000	$4,000	$4,000
Maximum payment	$35,000	$60,000	$100,000	$250,000
TOTAL WELLBORE LENGTH	GREATER THAN 6,000 FEET (MOSTLY UNCONVENTIONAL)			
Number of wells bonded	<26	26–50	51–150	>150
Flat fee	None	$140,000	$290,000	$430,000
Bond per additional well	$10,000	$10,000	$10,000	$10,000
Maximum payment	$140,000	$290,000	$430,000	$600,000

Source: Pennsylvania Department of Environmental Protection, n.d.

the number of wells it operates to anywhere between seventy-five and one hundred and fifty wells, it fulfills its entire bonding requirements for $60,000, with no additional cost per well.

While support for Act 13 was widespread among unconventional operators, small conventional operators were opposed to the regulatory changes. Conventional operators felt that their operations of fewer and shallower wells with a smaller wellsite footprint were less environmentally risky than unconventional operations. Thus, they saw themselves as unfairly grouped in with riskier unconventional operations and subject to overburdensome regulations that would affect the economic viability of their companies. Consequently, conventional operators continue to support proposals that would exempt them from certain regulatory requirements under Act 13.

Why Are Bonding Requirements Important?

Motivation for bonding requirements includes an efficiency justification and an equity justification. First, bonding requirements encourage firms to consider at least some of the cost of reclamation in the decision to drill a particular well. Incorporation of reclamation costs increases efficiency by preventing the drilling of wells that would be uneconomical if reclamation and bonding requirements were not in place. Second, when firms operating wells go bankrupt or dissolve, bonds provide funds for reclamation, reducing outlays for the state or affected landowners. Put differently, without bonds, greater costs would be shifted to parties that did not benefit economically from the well. The importance of bonding requirements—or some incentive for reclamation—is suggested by the large number of likely abandoned wells drilled in the decades following passage of Act 25 of 1955, which introduced requirements for reclamation but had no bonding requirements. According to state records, for about 5,400 wells drilled (or "completed") between 1955 and 1984, the DEP received no production data since it began collection of such information in 1980 and has no report of reclamation. The lack of production data and reclamation reports suggest that these wells are abandoned. These likely abandoned wells represent about 20 percent of all documented wells completed from 1955 to 1984. The high rate of abandonment underscores the importance of additional incentives for reclamation.

Appropriate bond amounts, however, may be as important as having any bonding requirements. Bonding requirements should reflect the cost of well reclamation. Otherwise, firms in financial stress or facing dissolution have more to gain by abandoning a well than by reclaiming it. If bonding requirements are below the cost of reclamation, reclaiming abandoned wells requires public funds, assuming that private landowners are unlikely to fund reclamation.

A Description of Well Reclamation

Once a well reaches the end of its economically useful life, the firm must reclaim it, which includes restoring the well site and plugging the well.

Well Site Restoration

The majority of well site restoration occurs immediately after the drilling of a well has been completed and the well begins producing. As set forth in the DEP's regulations, restoration includes regrading of any disturbed land to predrilling contours and uses, including restoration of vegetation; removal of, or filling of, pits used to contain produced fluids or industrial waste; implementation of a stormwater management plan to preserve the integrity of surface water environments as well as preventing accelerated erosion and sedimentation; and removal of all drilling equipment and infrastructure not needed for production (Pennsylvania Department of Environmental Protection 2012). To ensure that well site restoration accurately returns any disturbed land to predrilling conditions, the requisite stormwater management plan must include drawings and narratives describing the following: existing topographic features of the site and surrounding area; type, depth, slope, locations, and limitations of soils and geologic formations; past, present, and proposed land uses to the well site; and identification of all surface water features and description of the well site hydrogeology (Title 25 Pennsylvania Code §102.8). At the end of the well's producing life, the firm must plug the well and complete restoration of the site in accordance with the Clean Streams Law of Pennsylvania (58 Pa. Cons. Stat. §3216). Permit applications and accompanying documents require operators to detail, including providing maps and diagrams, their plans and ability to comply with well site restoration requirements. Operators that fail to abide by statutes, regulations, and the plans outlined in their permit application may be subject to fines and may lose the ability to operate until deficiencies are addressed.

Well Plugging

Well plugging consists of sealing off any rock formations that contain oil, gas, water, or brine with a cement or metal plug, called a mechanical plug in industry terms. The areas between the plugs are filled with a nonporous gel material, and a final cement plug caps the well at the surface. The plugs prevent liquids or gases in the surrounding subsurface from flowing into the well bore and to aquifers or the surface. Wells drilled in coal producing areas require an additional plug at least one hundred feet below the lowest workable coal seam to prevent the seam from flooding. With a few exceptions, the procedures for plugging conventional and unconventional wells are largely the same. The difficulty of plugging a conventional well depends on the condition of the casings that line the wellbore and the integrity of the wellbore itself. If the wellbore has collapsed, then it must be redrilled and cleaned out before it can be plugged. Because abandoned wells deteriorate over time, plugging costs are generally higher for older wells.

Plugging an unconventional well is more standardized because the wells are newer and are generally consistent in their casing condition and wellbore integrity. The vertical portion of an unconventional well is plugged much like a conventional well. The primary difference is that an unconventional well requires placement of a metal plug that is either removable or drillable under a cement plug close to the bottom of the vertical part of the well (about 20 feet above the horizontal portion of the well). The horizontal portion of the well is contained almost entirely within the producing shale formation and is not cemented. The plugs above it protect the shale from flooding as well as preventing stray methane from entering the vertical portion. Thus plugging an unconventional well where the distance from the surface to the bottom plug is one thousand feet is similar to plugging a one-thousand-foot conventional well.

How Much Does Well Reclamation Cost?

Conventional Wells

Surprisingly, there is little analysis of well reclamation costs. One exception is Andersen, Coupal, and White (2009), who study the cost of reclaiming abandoned wells in Wyoming and estimate the total outstanding reclamation costs for the state. The analysis looks at the costs incurred by the Wyoming Oil and Gas Conservation Commission for reclaiming 255 abandoned oil and gas wells in 48 locations from 1997 to 2007. The 255 wells were a mix of older and newer wells and had an average cost of reclaiming of $10.01 per foot of well depth, which translated to $27,555 for the average well. The relationship between total drilling depth of wells at any location was strong, with a correlation between depth and cost of 0.985. The strength of the correlation suggests that reasonable estimates of reclamation costs could come from a simple method where the cost per foot from plugged wells is applied to the depth of a given unplugged well.

Unconventional Wells

Over ten thousand shale gas wells have been drilled in Pennsylvania since 2006, and many more are projected over the coming years (Weber and Earle 2017). The profound depth of shale formations, which range from 5,000 to 9,000 feet in Pennsylvania, suggests that plugging unconventional wells will be more costly than plugging conventional wells. Despite the revising of bonding requirements under Act 13, the energy industry's history of busts and the number of unconventional wells that could be abandoned may give cause for concern. Only a handful of unconventional wells have been reclaimed, so there is little data on actual reclamation costs.

Because of the strong relationship between reclamation costs and well depth, a reasonable starting point for cost estimates is to use costs per depth measures from conventional wells and apply them to unconventional wells. This was the approach taken by Mitchell and Casman (2011), who estimated the cost of reclaiming unconventional wells in Pennsylvania. They summarize data from approximately one thousand well completion reports submitted to the DEP and show that the average total wellbore length for an unconventional shale gas well in Pennsylvania is approximately 10,675 feet. Using a rounded version of the Andersen, Coupal, and White (2009) dollar per foot cost of $10.50, Mitchell and Casman estimate the average reclamation cost for a single unconventional Marcellus Shale gas well to be about $110,000.

The estimate by Mitchell and Casman was based on reclamation costs of wells in Wyoming from 1997 to 2007 and can be improved in several ways.

First, the reclamation data are now several decades old, and labor and material costs could have changed. Second, Wyoming is not Pennsylvania. The two states have different economic contexts, meaning differences in prices for goods and services, as well as differences in geology and terrain, all of which could affect reclamation costs. Perhaps most importantly, Mitchel and Casman apply the cost per foot measure to the total length of an unconventional well, summing the vertical and horizontal portions. State regulation, however, only requires that portion from the surface to about twenty feet above the turning point in the well. Thus, the relevant depth is the distance from the surface to the top of the shale formation.

Data and Methods

To estimate the cost of reclaiming an unconventional gas well, we calculate the average cost per foot of reclaiming conventional oil and gas wells in Pennsylvania. We then apply various cost per foot measures to the average depth of permitted unconventional wells, which gives our estimate of reclamation costs.

We downloaded information from seventy-eight DEP well plugging contracts executed from 2005 to 2016 that were available through the state's contract library.[3] The contracts cover 1,205 wells, with an average contract size of fifteen wells but a range of one to one hundred thirty-eight. Most of the contracts have two parts, each covering distinct groups of wells. Part A includes the permit numbers, status, and well type of the plugged wells, an itemized schedule of unit prices for the equipment and materials, and a total cost for the wells covered in Part A. Part B includes the permit numbers, status, and well type of the plugged wells and only a lump sum cost for plugging the wells covered in Part B. The total cost of the contract is the combined amount of Parts A and B. Figure 11.1 is a sample contract. For all contracts, we put costs in 2010 dollars.

The variable of interest is each contract's average cost per foot plugged. To calculate well depth, we mapped wells over a spatial dataset displaying the producing oil and gas pools of Pennsylvania and their average production depths (Carter et al. 2015). Wells are assigned the depth of the pool from which they produced. Aggregating depth across all wells in the contract gives the contract's total depth. The average cost per foot is then calculated as the total contract cost (in 2010 dollars) divided by the total depth plugged. When averaging costs across contracts, we weight by the number of wells in the contract, which gives more weight to contracts with more wells.

To ensure that our depth measure is reliable, we look at a subset of 134 wells for which we had verified depth information. The 134 wells for which well depth

		SCHEDULE OF UNIT PRICES			
ITEM NO.	DESCRIPTION	ESTIMATED QUANTITY	UNIT	UNIT PRICE	ESTIMATED AMOUNT
	A. Schedule of Unit Prices for Cleaning Out and Plugging Four (4) Abandoned Gas Wells Nos. 049-00249, 049-00338, 049-00343 and 049-00344.				
1.	Site preparation, site restoration, restoration, and E&S controls.	Job	Job	Lump Sum	$4,227.28
2.	Mobilization/demobilization of basic plugging rig and related equipment.	Job	Job	Lump Sum	3,227.28
3.	Plugging rig operation time.	118	Hour	$165.00	19,470.00
4.	Stand-by time.	8	Hour	1.00	8.00
5.	Portland cement mixed and in place in wellbore.	340	944 Sack	12.00	4,080.00
6.	Bentonite powder mixed and in place in wellbore.	4	100v Sack	10.00	40.00
7.	Lost circulation material mixed and in place in wellbore.	4	Sack	10.00	40.00
8.	Material in surface hole.	Job	Job	Invoice	900.00
9.	Fresh water supply.	200	42 gal. Bbl.	1.50	300.00
10.	Waste fluid disposal cost & the treatment facility.	200	42 gal. Bbl.	0.75	150.00
11.	Fishing tool rental.	Job	Job	Invoice	3,000.00
12.	Trucking charge.				
	a. flat-bed track.	12	Hour	75.00	900.00
	b. Tractor-trailer.	8	Hour	75.00	600.00
	c. Water truck, disposal and supply hauling.	12	Hour	65.00	780.00
	Part A. Amount of Bid				$37,622.56
	B. Lump Sum Price for Well Nos. 049-00182, 049-06183, 049-00334, 049-00339 and 049-00342.				
	Lump Sum Price for Cleaning Out and Plugging Five (5) Abandoned Gas Wells.				
	Part B. Amount of Bid				$55,942.25
	Total Amount of Bid (Part A & Part B):				$93,564.81

FIGURE 11.1. Sample well plugging contract. Pennsylvania Department of Environmental Protection.

information is available had an average depth of 1,697 feet; for the same wells, the proxy variable for depth showed an average depth of 1,777 feet. The two measures of depth are therefore similar on average. Of course, they can diverge for a given well and, on average, the proxy depth deviated from actual depth by about 20 percent.

To explore how reclamation cost per foot varies, we consider average costs for small and large contracts and for contracts before and after shale development began. Contract size affects the cost of reclamation if there are economies of scale in reclamation, in which case plugging firms would be willing to make lower bids to obtain larger contracts. We therefore divide contracts into small and large groups based on whether the contract was smaller or larger than the median contract (six wells). The year of the contract may have also affected costs since firms that can reclaim wells can also provide drilling and well services for the shale industry. As shale development grew, firms may have found the state plugging contracts less attractive than the opportunities in private industry. We therefore break contracts into those that are preshale period contracts (2008 or before) or shale period contracts (after 2008).

Table 11.3 shows the number of contracts and wells by different contract types. The two periods had similar numbers of contracts (thirty-seven in 2008 or before and forty-one after 2008). Despite having slightly fewer contracts, the number of wells plugged in 2008 or before was much higher: 712 compared to 493. This indicates that contract sizes became smaller during the shale period.

In breaking out contract types, it became apparent that average cost per foot varied greatly with contract size. To further explore the relationship between contract size and average cost per foot, we regressed cost per foot on the contract size and the contract size squared. This permits estimating a nonlinear relationship between size and cost. The nonlinearity is likely; increasing contract size from one to two will probably have a different effect on average cost than increasing it from 99 to 100.

TABLE 11.3 State-reclaimed wells, by contract size and period, Pennsylvania

CONTRACT TYPE	NUMBER OF CONTRACTS	NUMBER OF WELLS
All	78	1,205
Small contract	39	113
Large contract	39	1,092
Preshale period	37	712
Shale period	41	493

Source: Pennsylvania Department of Environmental Protection. Well Plugging Contracts. See chapter 11, note 3. The contract library is available at https://contracts.patreasury.gov. Calculations by authors.

The average cost of plugging an unconventional gas well depends on the depth of the typical well. We used geologic data from Los Alamos National Labs to calculate the average depth of the Marcellus Shale for each county in Pennsylvania. We then calculate the average shale depth weighted by each county's number of permitted unconventional wells as of the end of 2017. This ensures that the average depth corresponds to the typical well, not the typical county.

Results

Across all contracts and wells, the average well had a reclamation cost of $7.7 per foot (figure 11.2). Because each contract's average cost is weighted by its number of wells, the average primarily reflects the costs of plugging wells in large contracts. This is because the average well belongs to a large contract. Among small contracts, those with six wells or fewer, the average cost was almost three times larger, at $20.3 per foot. The difference in cost between contracts from the preshale period and those in the shale period is less, $6.4 compared to $9.7. This could be interpreted as an increase in the cost of well services during the shale period. Contracts in the shale period, however, were also smaller, which have much greater average costs. Thus the difference in costs over time primarily reflects changes in the average contract size, not changes in the cost of services in the well plugging industry.

Figure 11.3 shows the relationship between contract size and average reclamation cost per foot, displaying actual size-cost pairs and the quadratic line that best fits the data as estimated by an ordinary least squares regression. Costs clearly decline with contract size over most contract sizes (one to fifty wells), with the rate of decline slowing. The estimated relationship between cost and contract size suggests that a one-well contract has an average cost of $23.9 per foot, but a fifteen-well contract costs $14.6 per foot. The figure also shows two contracts that have substantially higher average cost per foot than any other contract. Excluding them from estimation of the best-fit line gives fairly similar predictions: a one-well contract has an average cost of $19.3 per foot, and a fifteen-well contract costs $12.8 per foot.

The lower per foot cost of larger contracts undoubtedly reflects some degree of fixed costs being spread over more wells, such as the cost of mobilizing equipment to a general area. It may also reflect differences in the types of wells covered by small versus large contracts. Small contracts may be especially problematic wells, like those in a wetland or close to a house. Further analysis shows the opposite: larger contracts have a greater number of wells located in a wetland. However, larger-contracts areas are also more likely to be in rural areas, which would likely lower costs.

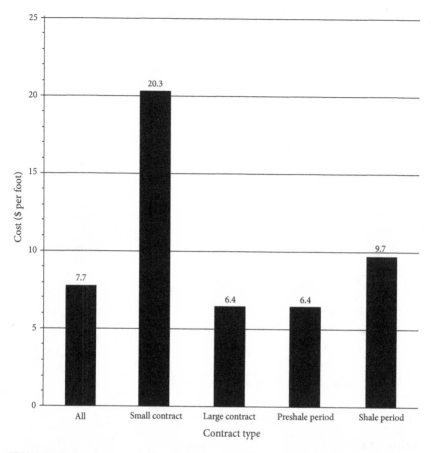

FIGURE 11.2. Average well reclamation cost per foot, by contract size and period. Calculations by authors from data from Pennsylvania Department of Environmental Protection, well plugging contracts; and Pennsylvania Department of Conservation and Natural Resources, well depth data.

Applying our cost estimates to unconventional wells requires specifying a contract size. It is hard to know the likely sizes of contracts to reclaim unconventional wells. Unconventional wells are located on pads with five or so wells, but pads are kilometers or more from one another. This is unlike the distribution of conventional wells, which are more scattered throughout the landscape, with the possibility of clusters in spots. A plausible scenario is that one unconventional well plugging contract would correspond to reclaiming all wells on one pad. Assuming five wells per pad means a contract size of five. On the high end, one could imagine three pads as falling under the same contract, which would include fifteen wells. At these contract sizes, the predicted cost per foot is $20.4 for five

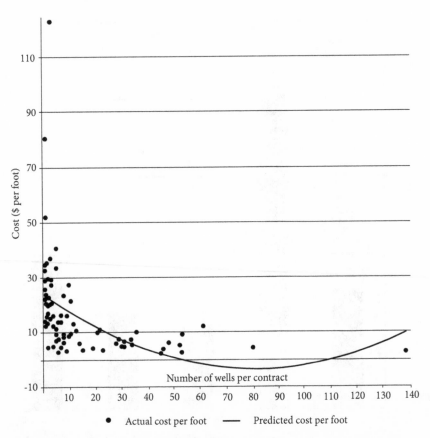

FIGURE 11.3. Predicted versus actual cost of reclamation. Calculations by authors from data from Pennsylvania Department of Environmental Protection, well plugging contracts; and Pennsylvania Department of Conservation and Natural Resources, well depth data. The estimated quadratic relationship between cost ($ per foot) and contract size (wells) is: Cost = 23.6 − 0.66 (wells) + 0.004 (wells²).

wells and $14.6 for fifteen wells. The average depth of unconventional wells in Pennsylvania is about 6,300 feet. Applying the cost per foot estimates gives a total cost of reclamation between $91,980 and $128,520.

Discussion

Caveats and Implications

The estimated range of plugging costs for unconventional wells is about ten times higher than current bonding requirements that apply to most unconventional wells, of $10,000 per well. This is in line with long-standing policy governing conventional wells where bond amounts are far below the cost of reclamation. Currently, conventional wells require a bond of $4,000 and we estimate the average plugging cost per well to be $45,000, given the average depth of active conventional wells at 3,100 feet and assuming a contract size of fifteen wells, which gives a per foot reclamation cost of $14.6.

In considering the quality of our estimate, one consideration suggests that the number might be too low, and another suggests the opposite. Our reclamation costs are based on conventional wells, which have a small surface footprint. Pads for unconventional wells can occupy ten acres because of the need to accommodate many trucks and tanks. As such, removing an unconventional pad and restoring the landscape to its prior contours will involve substantial earth moving and regrading along with the associated erosion control measures. Such costs are largely absent for conventional wells, so their costs exclude extensive site restoration work. On the other hand, unconventional wells are far more accessible than most conventional wells and their wellbores will be in better condition, both of which would make reclamation cheaper.

The only documented cost of reclaiming an unconventional shale well comes from a summary of Cabot Oil and Gas Corporation's Good Faith Efforts (2010) where they disclose paying over $2 million to plug three wells, a cost of about $700,000 per well. The number suggests much higher reclamation costs, but it should be viewed with caution. The DEP required Cabot to reclaim the wells in the wake of groundwater contamination events in Dimock Township, Pennsylvania. Because of the public salience of the case, Cabot may have had incentives to show high reclamation costs as an indication of its efforts to comply with state requirements.

Regardless of the exact estimate of reclamation costs, bond amounts and reclamation costs differ substantially, which raises the question of whether bond amounts need to be near reclamation costs to prevent abandonment. The state uses other tools to encourage reclamation, including levying fines and empowering the DEP to deny or withhold permits for continued or expanded operations for operators with outstanding reclamation obligations. However, we could not find evidence of any instances in which the DEP exercised its power to deny permits to operators. We are unaware of any research estimating the effect of bonding requirements on abandonment rates, but to the extent that contemporary wells are not abandoned en masse the combination of below-cost bond amounts and other tools seems to be working. Thus high bond requirements may not be needed to ensure reclamation in most instances.

While the current regulatory scheme may be effective in most cases, when a firm is exiting the market, below-cost bond amounts may incentivize the firm to abandon its wells, forfeit its bonds, and shift the remaining reclamation costs to the state and taxpayers. Such problematic incentives are particularly acute for blanket bonds. Currently, an unconventional operator with 150 wells can be bonded for $430,000, which is less than $3,000 per well or approximately 3 percent of our estimated unconventional well reclamation costs.

A state such as Pennsylvania that has a legacy of small "family-run" conventional operators may seek to keep bond amounts below reclamation costs in order to keep costs down for small operators. However, the existence of blanket bonds cuts against small operators, as blanket bonds inherently favor larger operators, or at least those with more wells. Additionally, a small operator justification of below-cost bond amounts serves little purpose for the unconventional industry, which is dominated by large operators.

Our estimate of reclamation costs may be limited in its application for informing efficient bond amounts for conventional and unconventional wells because our cost estimate is based on well reclamation that occurred decades after the wells were first abandoned. Reclamation costs increase as the time between abandonment and reclamation increases. Well deterioration over time makes it difficult to clean the well before it can be plugged, and the cost to mobilize equipment increases as access roads must be reopened. Accordingly, our estimate likely overstates the cost to reclaim a well at the time it no longer produces or shortly thereafter.

Given the current registration, bonding, and reporting requirements it seems unlikely that a massive number of unbonded conventional or unconventional wells will be abandoned and not reclaimed for decades as happened with conventional wells in Pennsylvania during the first century of the industry's development. A widespread bust in the shale gas industry could result in many hundreds or thousands of unconventional wells becoming the responsibility of the state, but the existence of those unconventional wells would be known and reclamation could commence without a long delay analogous to the state's contemporary effort to address abandoned conventional wells. Lacking an estimate based on firms' or the state's costs to reclaim wells shortly after abandonment, our estimate may serve as an upper bound on actual reclamation cost.

Alternatives to Bonds

Increasing bonding requirements is not the only way to pay for well reclamation when a firm abandons a well and cannot be brought to fund reclamation. Taxes on extraction could finance a well reclamation fund managed by a state or an independent entity that manages well reclamation contracts. In the case of Pennsylvania, an alternative to a new tax would be to increase the state's unconventional well impact fee and divert a portion of it into a reclamation fund.

Currently, the impact fee funds a Marcellus Legacy Fund. Despite the name, none of the money goes into an actual permanent fund. Rather, it is distributed to state agencies and local governments and can be spent in diverse ways. If just 10 percent of impact fee revenues were diverted into a reclamation fund, the fund could have $100 million in five years or less, which could fund reclamation of many wells. One could argue that such a fund should only pay for reclamation of unconventional wells. Otherwise, shale gas firms are paying for reclamation of wells abandoned by the conventional oil and gas industry in an era of lax enforcement by the state.

A problematic aspect of a tax-and-fund approach is that industry could interpret payment of the tax as shifting responsibility for reclamation to the state. If the transfer is formalized, then the state assumes full responsibility for reclaiming all wells. If the transfer is not formally recognized, law-abiding firms would pay for reclamation twice—once when they pay the tax and once when they reclaim their wells. A properly specified bond approach instead ensures that firms pay for reclamation once, either when they undertake reclamation or when they forfeit their bonds to the state.

Economic Efficiency of Plugging Abandoned Wells

It is probably not economically efficient to reclaim all abandoned wells. Some wells will be very costly to reclaim because they need to be redrilled before plugging or because roads need to be built to access them, or both. In such cases reclamation costs may exceed the value of the health and environmental risks that the wells pose. From an efficiency perspective, the state should only plug wells where the value of reclamation exceeds the cost. Moreover, since the DEP has a limited budget for plugging abandoned wells through the Orphan Well Plugging Fund, it should focus on wells where the benefits of reclamation exceed costs by the greatest amount.

On the other hand, equity considerations could motivate plugging wells even if reclamation costs exceed the benefits. Abandoned wells impose costs on nearby property owners or residents who may have not benefited from the well, and perhaps purchased the property assuming that the state would enforce its reclamation requirements. State-funded reclamation takes the burden of abandoned wells from a few nearby residents and distributes it across the state.

Final Remarks

Pennsylvania's governance of oil and gas extraction has evolved substantially over the last one hundred and fifty years, with policy innovations often occurring with major shifts in industry activity. The state's large inventory of abandoned oil and gas wells primarily reflects the inadequacy of earlier policies. About one hundred years of development occurred before the state began a permitting process needed to enforce reclamation requirements. About thirty years passed before bonding requirements emerged to provide additional incentive for reclamation. Bonding requirements for conventional wells have been—and continue to be—a fraction of reclamation costs as evidenced by our analysis of state well plugging contracts. The same applies to unconventional wells. The presence of civil penalties for well abandonment, however, may mean that bonding requirements need not equal reclamation costs in order to substantially reduce abandonment rates.

A separate issue is how to fund reclamation in instances where the firm responsible for a well cannot be found or brought to pay the cost of reclamation. Such cases are probably few and far between in eras of stable energy prices but could become common during a prolonged period of low prices that drive firms into bankruptcy. Although COVID-driven declines in oil prices have far exceeded natural gas price declines, a slow economic rebound could keep natural gas prices low for a while and drive some operators into bankruptcy. At the same time, state government resources to fund reclamation will likely fall as state tax revenues decline. This is clearly one value of having bond amounts equal to reclamation costs—they prevent the state and the taxpayer from having to pay for reclamation. Such requirements ensure that energy firms always pay the full cost of reclamation, either directly or indirectly by forfeiting their bonds to the state.

Notes

1. Footnote 60 on page 341 in Tarr and Clay 2015 mentions a statute from 1878 under Pa. Laws 57 concerning the plugging of oil wells, which may in fact be the first reclamation-related legislation passed in the state.

2. We compiled a database of the total number of unique wells contained in state records. Using a variable of the last year each well reported production to the DEP, we count the number of wells that were drilled under the policy and that had stopped reporting production in 2010 or before but that were not plugged by the end of the policy (2012).

3. The contract library is available at https://contracts.patreasury.gov. Oil and gas reclamation contracts can be found through using the "Subject matter contains" search capability.

References

Act 13. See 58 Pa. Cons. Stat. §2301 et. seq.

Andersen, Matthew A., Roger H. Coupal, and Bridgette White. 2009. "Reclamation Costs and Regulation of Oil and Gas Development with Application to Wyoming." *Western Economics Forum* 8:40–48. https://doi.org/10.22004/ag.econ.92846.

Boose, Kristian. 2011. "House Explosions in Bradford Pennsylvania Tied to Migrating Methane Gas from Drilling Activity." Protecting Our Waters, March 24. https://protectingourwaters.wordpress.com/2011/03/24/house-explosions-in-bradford-county-pennsylvania-tied-to-migrating-methane-gas-from-drilling-activity/.

Cabot Oil and Gas Corporation. 2010. "Exhibit B Summary of Cabot's Good Faith Efforts." September 28. http://www.jlcny.org/site/pdf/CB164695928.pdf.

Carter, Kristin M., Michael E. Moore, John A. Harper, Thomas G. Whitfield, Brian J. Dunst, Robin V. Anthony, Katherine W. Schmid, and April Chipman. 2015. "Oil and Gas Fields and Pools of Pennsylvania: 1859–2011." Pennsylvania Geological Survey, 4th ser., Open-File Report OFOG 15–01.0, 14 p., Portable Document Format (PDF), and associated File Geodatabase Feature Dataset.

Cone, Andrew, and Walter R. Johns. 1870. *Petrolia: A Brief History of the Pennsylvania Petroleum Region, Its Development, Growth, Resources, etc., from 1859 to 1869.* New York: D. Appleton. https://historicpittsburgh.org/islandora/object/pitt%3A00afh5490m.

Detrow, Scott. 2012a. "Perilous Pathways: How Drilling near an Abandoned Well Produced a Methane Geyser." StateImpact Pennsylvania, October 9. https://stateimpact.npr.org/pennsylvania/2012/10/09/perilous-pathways-how-drilling-near-an-abandoned-well-produced-a-methane-geyser/.

Detrow, Scott. 2012b. "Perilous Pathways: Abandoned Wells Don't Factor into Pennsylvania's Permitting Process." StateImpact Pennsylvania. October 12. https://stateimpact.npr.org/pennsylvania/2012/10/12/perilous-pathways-abandoned-wells-dont-factor-into-pennsylvanias-permitting-process/.

Dilmore, Robert M., James I. Sams III, Deborah Glosser, Kristin M. Carter, and Daniel J. Bain. 2015. "Spatial and Temporal Characteristics of Historical Oil and Gas Wells in Pennsylvania: Implications for New Shale Gas Resources." *Environmental Science and Technology* 49 (20): 12015–23. https://doi.org/10.1021/acs.est.5b00820.

Ely, Northcutt. 1933. *The Oil and Gas Conservation Statutes.* Washington, DC: US Federal Oil Conservation Board. https://books.google.com/books?id=ICoEAAAAMAAJ&printsec=frontcover&source=gbs_ge_summary_r&cad=0-v=onepage&q&f=false.

Kang, Mary, Shanna Christian, Michael A. Celia, Denise L. Mauzerall, Markus Bill, Alana R. Miller, Yuheng Chen, Mark E. Conrad, Thomas H. Darrah, and Robert B. Jackson. 2016. "Identification and Characterization of High Methane-Emitting Abandoned Oil and Gas Wells." *Proceedings of the National Academy of Sciences* 113 (48): 13636–41. https://doi.org/10.1073/pnas.1605913113.

Kang, Mary, Cynthia M. Kanno, Matthew C. Reid, Xin Zhang, Denise L. Mauzerall, Michael A. Celia, Yuheng Chen, and Tullis C. Onstott. 2014. "Direct Measurements of Methane Emissions from Abandoned Oil and Gas Wells in Pennsylvania." *Proceedings of the National Academy of Sciences* 111 (51): 18173–77. https://doi.org/10.1073/pnas.1408315111.

Litvak, Anya. 2014. "Stranded Methane Gas in Versailles a Century in the Making." *Pittsburgh Post-Gazette*, June 14. http://www.post-gazette.com/powersource/consumers-powersource/2014/06/15/Stranded-methane-gas-in-Versailles-a-century-in-the-making/stories/201406150067.

Mitchell, Austin L., and Elizabeth A. Casman. 2011. "Economic Incentives and Regulatory Framework for Shale Gas Well Site Reclamation in Pennsylvania." *Environmental Science and Technology* 45 (22): 9506–14. https://doi.org/10.1021/es2021796.

58 Pa. Cons. Stat. §2301 et. seq. (Act 13)

Title 25 Pa. Code §102.8.Pennsylvania Department of Environmental Protection. 2000. "Pennsylvania's Plan for Addressing Problem Abandoned Wells and Orphaned Wells." Bureau of Oil and Gas Management. Doc# 550–0800–001, April 10.

Pennsylvania Department of Environmental Protection. 2012. "Oil and Gas Industry Training. Section 2: Site Restorations and Post-Construction Stormwater Management Plans (PCSMs)." Office of Oil and Gas Management, July 6. http://files.dep. state.pa.us/OilGas/BOGM/BOGMPortalFiles/OilGasReports/2012/Training_ Materials/SiteRestoration2012(2x2Version).pdf.

Pennsylvania Department of Environmental Protection. 2019. "2019 Oil and Gas Annual Report." https://storymaps.arcgis.com/stories/3f99825a393d4fe080d6d 1c8e74b6f34.

Pennsylvania Department of Environmental Protection. n.d. Act 13: Frequently Asked Questions. https://www.dep.pa.gov/Business/Energy/OilandGasPrograms/ Act13/Pages/Act-13-FAQ.aspx.

Shen, Jialei, and Zhi Gao. 2018. "Ozone Removal on Building Material Surface: A Literature Review." *Building and Environment* 134:205–17. https://doi.org/10.1016/j. buildenv.2018.02.046.

Skinner, John F. 1990. *A History of New York State Natural Gas Corporation 1913–1965.* Clarksburg, WV: CNG Transmission Corporation.

Tarr, Joel A., and Karen Clay. 2015. "Boom and Bust in Pittsburgh Natural Gas History: Development, Policy, and Environmental Effects, 1878–1920." *Pennsylvania Magazine of History and Biography* 139 (3): 323–42. https://www.jstor.org/ stable/10.5215/pennmaghistbio.139.3.0323?seq=1.

Townsend-Small, Amy, Thomas W. Ferrara, David R. Lyon, Anastasia E. Fries, and Brian K. Lamb. 2016. "Emissions of Coalbed and Natural Gas Methane from Abandoned Oil and Gas Wells in the United States." *Geophysical Research Letters* 43 (5). https://doi.org/10.1002/2015GL067623.

Waples, David A. 2014. *The Natural Gas Industry in Appalachia: A History from the First Discovery to the Tapping of the Marcellus Shale.* Jefferson, NC: MacFarland.

Weber, Jeremy G., and Andrew Earle. 2017. "Abandoned Unconventional Natural Gas Wells: A Looming Problem for Pennsylvania?" *University of Pittsburgh Shale Gas Governance Center Energy and Environment Blog,* January. https://medium. com/@GSPIAe_eBlog/abandoned-unconventional-natural-gas-wells-a-looming-problem-for-pennsylvania-ce8243aaf8a4.

PRIVATE AND PUBLIC ECONOMIC IMPACTS OF FRACKING IN WYOMING

Gavin Roberts and Sandeep Kumar Rangaraju

During the dry years, the people forgot about the rich years, and when the wet years returned, they lost all memory of the dry years. It was always that way.

—John Steinbeck, *East of Eden*

Wyoming provides a unique example of the emergence of hydraulic fracturing technology, and the associated economic consequences. It is one of the largest energy-producing states in the United States and has experienced its fair share of the boom and bust cycles associated with fossil fuel production. Before the shale gas revolution, the Rocky Mountain region, and especially Wyoming, was considered an extremely important source of future supplies of increasingly scarce natural gas in the United States. Another important source would be the liquefied natural gas import facilities that are now being converted to export facilities. Important technological advancements associated with fracking in Wyoming played a role in unleashing the shale gas revolution. These advancements led to a short-lived boom in gas production in Wyoming. Ironically, these technologies would be applied at lower cost in other states closer to the largest markets eventually decreasing the relative importance of Wyoming's newly unlocked gas reserves.

Fracking of both tight sandstone reservoirs and tight shale reservoirs boomed in Wyoming between 2005 and 2011, as real natural gas prices climbed to historically high levels, and the continued availability of "sweet spots" in Wyoming kept producers drilling even after prices fell precipitously in 2008 following the financial crisis. However, as gas prices remained depressed in the early 2010s culminating in the Opal natural gas price (Wyoming's benchmark natural gas price) falling below \$2.00 per MMBTU in 2012, producers reacted by reallocating drilling capital to more productive prospects outside the state.

High natural gas prices accompanied Wyoming's gas production boom during 2005 through 2008 and the combination of high production and prices was a boon to the state's coffers via the state's severance tax on the value of production. The state government expanded its workforce by over 10 percent between 2005 and 2010, but dramatic decreases in gas prices and production reversed Wyoming's fortunes. The boom also benefited residential, industrial, and commercial natural gas consumers. Wyoming is the largest energy consumer per capita in the United States after Louisiana, so low energy prices have a relatively large consumer surplus effect in the state.[1] Weather-related residential consumption has remained consistent with historical patterns in Wyoming since the new era of low natural gas prices began in 2009, but industrial and commercial consumption both grew by more than 40 percent between 2000 and 2017.[2]

Concerns that total benefits would be hampered as a result of externalities associated with fracking led Wyoming to adopt fracking regulations relatively early on. Wyoming has some of the most stringent regulations associated with hydraulic fracturing, and was one of the first states to require disclosure of fracking fluids (Bleizeffer 2010). While consumers in Wyoming continue to benefit from low natural gas prices, and a well-regulated natural gas industry, the primary threat facing the state is decreasing government revenues from severance taxes—the most important and variable component of tax revenues in Wyoming.

This chapter will provide the historical context of hydraulic fracturing, and will analyze the economic impacts of shale fracking, and how those impacts differed between the private and public sectors in Wyoming using a variety of data sources. The chapter will also propose policies to make the public sector less susceptible to resource-related short-term boom and bust cycles and protect the private sector from the associated spillover.

The History of Fracking in Wyoming

Wyoming represents a unique untold story in the emergence of hydraulic fracturing in the United States that has led to the shale gas revolution. Wyoming played a key role in the emergence of fracking technologies. It was one of the locations where massive hydraulic fracturing occurred in the 1970s in which the United States Department of Energy used nuclear explosions to fracture tight sandstone reservoirs (Law and Spencer 1993). In September 1992, McMurry Oil Company unlocked the massive natural gas reserves in the Jonah Field in southwest Wyoming by successfully fracturing the tight sandstone reservoirs found there (Noble 2014). This was more than five years before Mitchell

Energy would successfully apply similar techniques to the tight shale source rock in the Barnett Shale in Texas (Gold 2014). The relatively early onset of hydraulic fracturing in Wyoming gave the state a head start regulating fracking. Further, while the shale revolution has been primarily defined by the combination of horizontal drilling and hydraulic fracturing in the United States, Wyoming's unique geology has led to a boom in the combination of directional drilling and fracking.[3]

Wyoming was the fifth largest natural gas producing state in the United States in 2016 but fell to seventh in 2017 as it was overtaken by Louisiana and Ohio. Natural gas production boomed in Wyoming between 2000 and 2010 when gross natural gas withdrawals increased by 100 percent, but has steadily declined since 2010, falling approximately 30 percent.[4] This boom in Wyoming natural gas production primarily occurred in Sublette County, Wyoming, in the

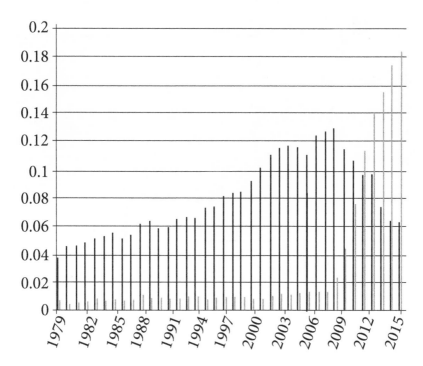

■ Wyoming proportion of US reserves

▨ Pennsylvania proportion of US reserves

FIGURE 12.1.   Proportion of US total natural gas reserves in Wyoming and Pennsylvania. Energy Information Administration. https://www.eia.gov/naturalgas/data.php#exploration.

Jonah Field and the Pinedale Anticline, where the combination of new hydraulic fracturing technology and new pipeline capacity led to a boom in drilling (Noble 2014). At the height of the boom in 2008 and 2009, Wyoming's natural gas reserves accounted for more than 10 percent of US total natural gas reserves. However, when natural gas prices declined after the financial crisis, and during the subsequent nationwide boom in shale production, many of Wyoming's relatively expensive natural gas drilling prospects became uneconomical, while economical reserve estimates continued to grow in the new shale giants such as Pennsylvania (figure 12.1).

Increasing natural gas production in the first decade of the twenty-first century was a boon to Wyoming's economy, but declining production since 2010 has led to a financial crisis in the state government, especially in the public education system. Tax revenues in Wyoming are heavily dependent on severance taxes

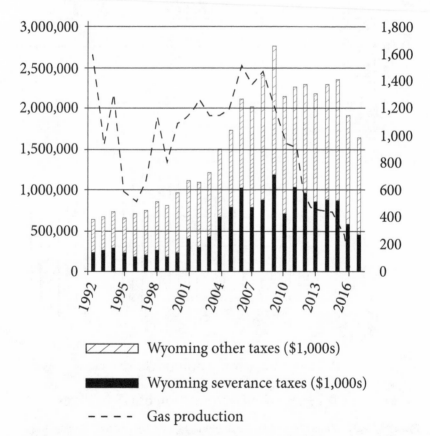

FIGURE 12.2. Severance and other tax revenues in Wyoming. US Census Bureau. https://www.census.gov/programs-surveys/stc.html.

associated with fossil fuel mining. In fact, variation in severance taxes resulting from variation in commodity prices and production caused more than 90 percent of the year-to-year variation in total tax revenues in Wyoming between 1992 and 2017. Figure 12.2 illustrates that severance tax revenues and total tax revenues peaked around the same time as natural gas production and drilling in 2009 when natural gas prices were also relatively high.

The Economic Impacts of Fracking in Wyoming

Historically, very few regions where exhaustible resource extraction makes up a significant proportion of economic output have been able to escape the associated notorious boom and bust cycles, especially in the case of petroleum extraction. The underlying causes and magnitudes of boom and bust cycles, however, vary across geographic location. For example, boom and bust cycles in Saudi Arabia are often the result of poor governance (Aguilera and Radetzki 2016, 42–58), while boom and bust cycles in the United States are more likely to be the result of global price volatility, or new discoveries of less costly or more productive drilling prospects in other locations.

The economic impacts of the shale gas revolution, or more generally the emergence of a new technology that impacts exhaustible resource extraction, will also vary by geographic location for a variety of reasons. Space-heating consumers of natural gas in an area with long harsh winters will benefit more from increased supplies and lower prices than space-heating consumers in an area with relatively mild winters. Wyoming consumers have benefited and will continue to benefit from lower natural gas prices for this reason. Benefits associated with hydraulic fracturing to producers will also vary by location primarily for geological reasons. Producers in an area with a large conventional natural gas resource base such as the Gulf of Mexico might be hurt as new supplies of shale gas decrease prices and make investments in major long-term projects uneconomical. On the other hand, an area with a large unconventional resource base that can be accessed through fracking such as the Marcellus Shale in Pennsylvania might experience a boom. Wyoming is uniquely situated between these two extremes: both conventional and unconventional natural gas reserves are abundant in the state (WSGS 2018).

The emergence of hydraulic fracturing and directional drilling unlocked gas reserves in Wyoming (primarily in tight sandstone reservoirs), which was beneficial for the state. Further, new technologies were applied relatively early in Wyoming, which allowed producers to cash in on high natural gas prices before 2008. However, when the shale revolution took hold, and natural gas prices fell across the entire United States, many relatively expensive drilling prospects in Wyoming became uneconomical at lower natural gas prices. This phenomenon is made clear by reviewing reserve numbers across locations. For example, natural gas reserves continued to grow in Pennsylvania during the early 2010s as per unit drilling costs in the Marcellus continued to decrease even as natural gas prices decreased, while Wyoming natural gas prospects became uneconomical and reserve estimates subsequently decreased (figure 12.1). The process by which natural gas producers only drill the most productive, or least costly, wells during periods of low prices, and expand to less productive, or costlier, drilling prospects during periods of high prices is termed endogenous well selection (Mason and Roberts 2018). Endogenous well selection will exacerbate boom and bust cycles in regions with many drilling prospects near the margin of profitability, such as Wyoming.

Clearly, the economic impacts of a new technology applied to an exhaustible resource are multiple and complex. We do not attempt to account in this chapter for all the potential economic impacts of hydraulic fracturing in Wyoming but give a broad overview of the impacts of changes in natural gas prices, drilling, and production on Wyoming's public and private sectors. For private-sector impacts of shale gas in Wyoming, we estimate the impact of natural gas prices, drilling, and production on private-sector job creation, while for the public sector, we focus on the impact of prices, drilling, and production on severance tax revenues. Job creation measures new jobs at expanding establishments and new jobs at opening establishments in the private sector and is collected from the Business Employment Dynamics database (BED) from the US Bureau of Labor Statistics (BLS 2019). Job creation data captures microeconomic adjustments in the private economy and has been used to measure the effects of oil price changes in different sectors of the economy (Davis and Haltiwanger 2001; Davis, Haltiwanger, and Schuh 1998).

Private-Sector Economic Impacts

Between 2002 and 2011, there were large increases in labor and capital used to extract natural gas from tight reservoirs in Wyoming, as indicated by the boom in the number of directionally drilled wells shown in figure 12.3. However, the number of natural gas wells being drilled in Wyoming decreased dramatically after 2011 because of lower natural gas prices and more profitable drilling prospects in other regions—a manifestation of the endogenous well selection mentioned above and also shown in figure 12.3. Further, the number of unique oil and gas well operators peaked at the same time as the number of wells and production. As

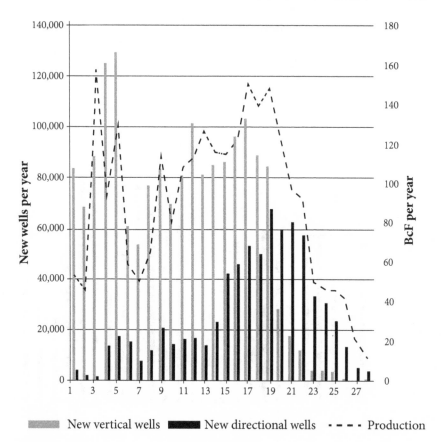

FIGURE 12.3. New natural gas wells and production in eight largest gas producing counties in Wyoming, 1990 to 2017. Energy Information Administration. https://www.eia.gov/dnav/ng/hist/n9010wy2a.htm.

drilling began to decline, the number of unique operators declined as well, indicating firms were either going out of business, leaving the state, or consolidating through mergers, as shown in figure 12.4.

Figure 12.5 indicates that job creation outpaced job destruction for most of Wyoming's hydraulic-fracturing and directional-drilling boom between 2002 and 2011, indicating economic expansion. The only exception occurred during 2009 in the depth of the "great recession." The total number of private-sector jobs created during the shale boom peaked at 24,237 in the first quarter of 2008, while job destruction in Wyoming peaked at 25,355 in the second quarter of 2009 in the middle of the great recession. The shale boom, therefore, appears to have offered a cushion to the Wyoming economy just before the effects of the recession were felt. We can observe the decline in job flows during the global financial

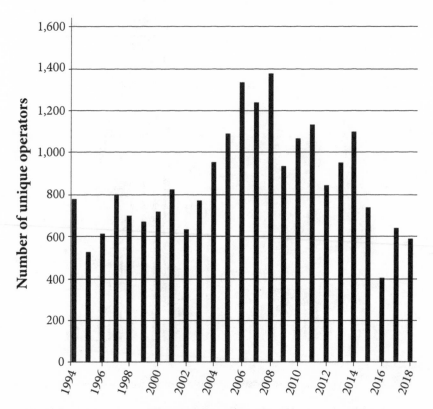

FIGURE 12.4. Unique oil and gas well operators, Wyoming, 1994–2018. DrillingInfo. https://app.drillinginfo.com/production.

crisis and it coincides with the collapse of natural gas prices, oil prices, and coal prices, which was induced by lower global demand for these commodities. Oil prices recovered relatively quickly after the financial crisis before the tight oil revolution began to take hold in 2014, while natural gas prices remained low after the financial crisis as a result of new supplies from shale hitting the market. In order to quantify the effect of natural gas prices, drilling, and production on Wyoming's private economy, we run several regressions with state-level private-sector job creation as the dependent variable.

Job creation data from the BED is available on a quarterly basis between the third quarter of 1992 and the third quarter of 2017 providing 101 observations for our analysis (US Bureau of Labor Statistics 2019). We use average quarterly natural gas prompt month prices (one-month futures prices) collected from the EIA as our measure of natural gas prices. The majority of natural gas deliveries in

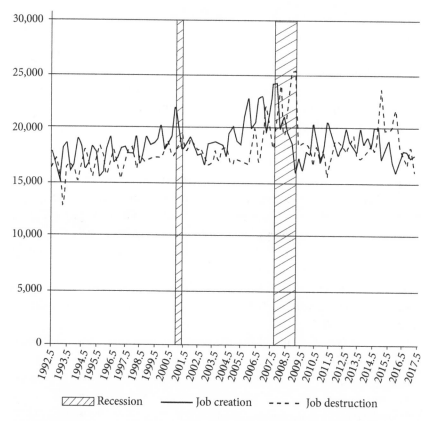

FIGURE 12.5. Job creation and job destruction in Wyoming. United States Census Bureau. Business Dynamics Statistics. 2017. https://www.census.gov/programs-surveys/bds.html.

the United States are priced as a function of the prompt month settlement price (Kaminski 2012), which makes variation in the prompt month price the best measure of variation in the price actually received for natural gas delivered to consumers. These prices are available from the US Energy Information Administration (EIA) website starting in 1994, so regressions that include prices have ninety-five quarterly observations.[5] Natural gas production and drilling data are constructed using data collected from the Wyoming Oil and Gas Conservation Commission website.[6]

The results of our private-sector economic impact analysis are displayed in table 12.1. We regress private-sector job creation on the covariates of interest. The first column of table 12.1, labeled "All Independent," reports the coefficient estimates for a regression with all three independent variables of interest, which are drilling (number of wells drilled in the quarter), production (total natural gas produced in a quarter in billions of cubic feet [Bcf] in the quarter), and prompt month prices (average closing price in dollars per MMBTU in the quarter). The remaining three columns show regressions with private-sector job creation as the dependent variable regressed on each independent variable individually. We chose to regress job creation on each independent variable individually due to relatively high correlation between the independent variables, especially drilling and production, while the regression with all three variables is necessary to check for the potential of omitted variable bias in the univariate regressions especially in light of correlation between these variables. The correlation between drilling and production is large with a correlation coefficient of 0.78, which is unsurprising as drilling is the mechanism that leads to increased production. Also, unsurprisingly, higher prices are positively correlated with drilling and production in Wyoming (a price taker given the size of the natural gas market in the United States), with correlation coefficients of 0.52 and 0.44, respectively. Examining the results from the individual regressions will help us understand the effects of these correlations in the full regression. We also ran each regression with lagged job creation and a time trend, but the results remained consistent, so we do not present those results here.

The full regression, with all three independent variables, indicates that an additional well drilled in a quarter will be associated with 0.018 additional jobs created in that quarter, but this result is statistically insignificant. The full regression indicates that a one Bcf increase in natural gas production in a quarter actually decreases job creation by 0.35 jobs, but this estimate is also statistically insignificant. The negative coefficient on the production variable is likely the result of collinearity with drilling and prices, as we discuss below when interpreting the results of the univariate regressions. Finally, the estimated coefficient on the prompt month futures price is large and statistically significant. A one-dollar increase in the US benchmark (Henry Hub) prompt month futures price

TABLE 12.1   Results of private-sector economic impact analysis

VARIABLE	ALL INDEPENDENT	DRILLING	PRODUCTION	PRICE	PRICE(UT)	PRICE(SD)
Drilling	0.0180	0.0405***	.	.	.	.
Production	−0.3519	.	3.9378***	.	.	.
Price	127.77***	.	.	141.72***	177.39*	−6.006
Constant	16768***	17675***	17690***	16907***	69734***	21653***
R^2	0.32	0.08	0.08	0.31	0.04	0.00
N	95	101	101	95	95	95

*** Significant at the 1% level.

** Significant at the 5% level.

* Significant at the 10% level.

is associated with approximately 128 additional jobs being created in a quarter.[7] The sample standard deviation of our quarterly price series is 6.74, so a one standard deviation increase in the prompt month futures prices is associated with more than 860 additional jobs being created in a quarter—an economically significant impact in a state with fewer than 300,000 jobs (Wen 2018). Clearly, natural gas prices have the largest impact on private-sector job creation in the state of Wyoming out of the three independent variables in our analysis.

The univariate regression with drilling as the independent variable indicates a statistically significant coefficient that is almost twice as large as the coefficient in the full regression of 0.041. By this estimate, a one standard deviation increase in drilling leads to almost five hundred additional jobs created in Wyoming in a quarter. The change in the magnitude of this estimated coefficient almost surely results from the strong correlation between drilling and production. Similarly, the estimated coefficient in the univariate regression with just natural gas production actually changes signs. The coefficient on production in the univariate regression is a positive and statistically significant point estimate of 3.94. This indicates that a one standard deviation increase in production will be associated with 480 additional jobs being created in a quarter. Finally, the estimated coefficient on price in the univariate model continues to be statistically significant and is slightly larger than in the full model. The univariate estimate indicates that a one standard deviation increase in price will be associated with more than 950 additional jobs created in a quarter.

It is important to note that while the drilling and production coefficients are statistically significant in the univariate regressions, the R^2 value in both of these regressions is just 0.08, while the R^2 value in the full regression is 0.32 and in the univariate price regression is 0.31. In fact, the adjusted R^2 is larger in the univariate price regression than in the full regression indicating that the inclusion of the drilling and production variables adds very little explanatory power to the model. Intuitively, changes in price are driving changes in drilling and production, and changes in price alone achieves the necessary explanatory power to pick up the impacts of drilling and production.

In order to provide some counterfactual context to our analysis we have also included the results of univariate regressions of job creation on natural gas prices in the US states of Utah and South Dakota. The results of these regressions are presented in the last two columns of table 12.1. The estimated coefficient for Utah is positive and statistically significant and is actually larger in magnitude than the coefficient in Wyoming. However, the R^2 value in the Utah regression is just 0.04 indicating that variation in natural gas prices is associated with very little variation in job creation in Utah. The estimated coefficient for South Dakota is actually negative, but statistically insignificant. These estimates highlight the degree to which private-sector job creation in Wyoming is dependent on natural gas prices relative to similar states where natural gas production makes up a much smaller proportion of economic output.

We also investigated the possibility that the impact of natural gas prices on private-sector job creation has changed over time. We ran the analogous univariate regression of job creation on price for three time periods: preshale era (1994 to 2005), postshale era (2005 to 2017), and post–natural gas price collapse (2009 to 2017). The coefficients associated with these regressions were 71.72**, 157.16***, and 46.62, where the latter estimate was statistically insignificant. The preshale and postshale coefficients are consistent with the full sample results, although job creation dependence on natural gas prices appears to have increased during the postshale era to some degree. This likely results from the increasing importance of the natural gas industry in Wyoming's economy after the shale boom began to accelerate. On the other hand, the relatively small and statistically insignificant post–price crash estimate indicates that there may be some asymmetry to the effect of natural gas prices on job creation in Wyoming—when natural gas prices increase, more jobs are added than jobs are lost when natural gas prices decrease.

Another private-sector impact of natural gas production in Wyoming comes from royalties paid to private mineral owners. One recent paper calculates that the average royalty rate earned by mineral owners in Wyoming is approximately 15 percent (Brown, Fitzgerald, and Weber 2016). Applying this rate to our price and production data, we generated estimates of monthly natural gas royalties paid to private mineral owners per month in Wyoming. A time series of these estimates is displayed in figure 12.5. Unsurprisingly, royalties roughly track production and prices, as displayed in figure 12.6, and are thus subject to the boom and bust cycles common in natural resource extraction. Recent research has indicated that natural gas royalties have a multiplier effect of approximately 1.5, so $1.00 in additional royalties will add approximately $1.50 in total income in Wyoming (Brown, Fitzgerald, and Weber 2017).

We conclude this section by emphasizing that natural gas prices are the primary driver of the economic impact of natural gas on the private sector in Wyoming. The quantitative findings are consistent with our introductory discussion of potential economic impacts of the shale revolution in Wyoming. Hydraulic fracturing combined with directional drilling to unlock natural gas from tight sandstone reservoirs occurred in Wyoming early relative to similar technological advances that eventually drastically increased natural gas production in other regions of the United States. This allowed producers in Wyoming to capture the high natural gas prices that predominated before 2009, and to expand their businesses. This expansion of drilling and production likely had spillover effects into

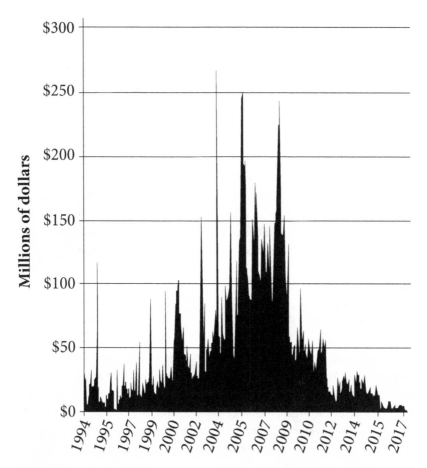

FIGURE 12.6.   Estimated monthly natural gas royalties paid to private mineral owners, Wyoming, 1994 to 2017. Energy Information Administration; Brown, Fitzgerald, and Weber (2016).

Note: Excludes lease bonuses.

the wider Wyoming economy via both job creation and the royalties multiplier effect, and also led to large severance tax revenues for the state.[8] However, when prices decreased due to larger aggregate supply of natural gas, drilling and production decreased in Wyoming, and the gas that was produced could only be sold into the new low price environment leading to a double whammy for Wyoming's natural gas producers, wider private economy, and state. Contrarily, in a state like Pennsylvania, with many extremely productive drilling prospects and decreasing costs related to efficiency improvements, gas must be sold at a lower price, but production and drilling remained robust throughout the 2010s and continue to increase at present.

Public-Sector Economic Impacts

The combined effects of lower prices and lower production after 2011 have seriously affected Wyoming's public finances, which depend heavily on severance tax revenues. Severance tax revenues accounted for approximately 40 percent of total tax revenues in Wyoming on average between 1994 and 2016, and as mentioned in the introduction, variation in severance taxes accounted for more than 90 percent of variation in total taxes in Wyoming over that time period.[9] For purposes of comparison, consider that the analogous proportions in Utah and South Dakota are just 1.4 percent and 0.5 percent, respectively, while variation in severance taxes were associated with approximately 70 percent and 30 percent of variation in total taxes in Utah and South Dakota.[10] Wyoming's state government expanded dramatically during the shale boom, as indicated by increased government employment. Government employment in Wyoming increased more than 10 percent between 2005 and 2011 in Wyoming from just over sixty-five thousand employees to more than seventy-two thousand employees. The state government is currently attempting to cut back on spending, but path dependence in government spending associated with "use it or lose it" budget incentives is making the task more difficult.

Now, we explore the quantitative effects of natural gas prices, drilling, and production on Wyoming's state budget to determine to what extent the state government of Wyoming depends on the natural gas industry. In this section, we present results from analogous regressions to those presented in table 12.1, but here we use severance tax revenues measured in thousands of dollars per year as the dependent variable of interest. In this simple analysis, we do not model the responses of firms to severance taxes, but firm responses to tax rates might play an important role. Natural gas producing firms are expected to extract at lower rates as a result of severance taxes (Lockner 1964). However, empirical work has shown that the primary result of lowering severance tax rates is to reduce state revenues but has little effect on firm behavior (Kunce 2003). This latter finding means that reverse causality from severance tax revenues to drilling and production outcomes should not present a problem in our empirical framework.

The data for our public-sector economic impact analysis include the same set of independent variables that we used in the private-sector analysis. We collected data on severance tax revenues in Wyoming from the Annual Survey of State Government Tax Collections (STC) from the United States Census Bureau website.[11] Unfortunately, these data are only available on an annual basis, so we converted our three independent variables to annual aggregates (annual average for the price series), which unfortunately drastically decreased the number of observations available.

Table 12.2 shows the results of our public-sector analysis. The first column shows results from a regression of annual severance tax revenues on annual natural gas prices, drilling, and production. The coefficient estimates on all three independent variables are statistically significant in the multiple regression. Interestingly, when controlling for production and price, higher levels of drilling are actually associated with lower severance tax revenues. The sample standard deviation of wells drilled per year between 1992 and 2016 is approximately 44,000, so the coefficient estimate indicates that a one standard deviation increase in wells drilled in a year will decrease severance tax revenues by almost $500 million. The negative coefficient on drilling is counterintuitive, but likely partially results from the high degree of collinearity between drilling and production ($r = 0.85$), and between drilling and prices ($r = 0.60$) in the annual data set—higher production and prices are associated with higher drilling. Another factor driving the large negative coefficient on drilling may be the fact that between 1994 and 2003, there were tax incentives associated with drilling new wells in Wyoming—wells drilled during that period would pay a severance tax of just 2 percent during their first twenty-four months of production versus the regular 6 percent severance tax rate on natural gas production (WYDOR 2010). Therefore, severance tax revenues could decrease to the extent that production from new wells began to account for an increasing proportion of production during that period, and production from wells drilled before 1994 declined. Indeed, when we ran the multiple regression with an indicator for the years without this new-well tax incentive, we found a coefficient on drilling that was insignificantly different from zero. The coefficient on drilling in the univariate regression is small and statistically insignificant.

State-level tax incentives can play a major role in natural gas drilling decisions, and states must balance the benefit of revenues associated with higher tax rates against potential losses in investment associated with higher rates. The

TABLE 12.2 Results of public-sector economic impact analysis

VARIABLE	ALL INDEPENDENT	DRILLING	PRODUCTION	PRICE	PRICE (UT)	PRICE (SD)
Drilling	−9.8846***	−0.9205	.	.	.	.
Production	725.71*	.	5.8200	.	.	.
Price	120625***	.	.	76224.4**	7778.3*	−403.68
Constant	318150**	672392***	582029***	16907***	34561*	7745***
R²	0.56	0.01	0.00	0.24	0.15	0.07
N	23	25	25	25	25	25

*** Significant at the 1% level.
** Significant at the 5% level.
* Significant at the 10% level.

complexities of optimal state-level severance tax strategies are outside the scope of this chapter, but Maniloff and Manning (2018) provide a theoretical model and empirical evidence for the interested reader.

The coefficient on production is positive and significant at the 10 percent level in the full regression. The sample standard deviation of production in the annual data set is approximately 390 Bcf per year, so the coefficient on production in the full regression indicates that a one standard deviation increase in production increases severance tax revenues by about $285 million. The coefficient on production in the univariate regression is small and statistically insignificant. It is interesting to note that the impacts of drilling and production are only revealed when controlling for price. Similar to our private-sector analysis above, the fact that drilling and production add very little explanatory power to the model leads us to focus on the role of prices. Again, the regression with all three independent variables is important to consider due to the risk of omitted variable bias in the univariate regression especially in the context of large correlation between prices, drilling, and production.

Natural gas prices have the most significant impact and most explanatory power on severance tax revenues out of the three variables of interest. The coefficient on prices is positive, large, and statistically significant in both the multiple regression and univariate regression. The multivariate regression coefficient indicates that just a one-dollar increase in annual average natural gas futures price will increase severance tax revenues by more than $120 million, while the univariate regression coefficient indicates a one-dollar increase in the futures price will increase severance tax revenues by more than $75 million. The standard deviation of the futures price is more than $2.00 in the annual sample, so a one standard deviation increase in the futures price is estimated to increase tax revenues by more than twice these amounts. The estimated coefficient on price in the univariate regression is likely biased toward zero as a result of the omitted variable bias mentioned above.

Note that the R^2 values in the drilling and production univariate regressions are extremely small. This indicates that these variables individually have very little explanatory power for severance tax revenues. However, these variables more than double the R^2 value when they are added to the price regression, from 0.24 to 0.56, indicating that they provide important explanatory power given that price is controlled for. This observation can be contrasted with the private-sector analysis in which drilling and production had very little explanatory power in their respective univariate regressions and did not significantly increase explanatory power in the price regression. Therefore, drilling and production did not have a significant impact on private-sector job creation over our sample period, but given a certain price level, drilling and production had significant impacts on severance tax revenues.

It is important to note that while drilling in the current period may be associated with lower severance tax revenues, and therefore less drilling in the current period has historically been an indicator of higher severance tax revenues in the current period, drilling is the mechanism that leads to more future production in gas and oil production (Mason and Roberts 2018; Anderson, Kellogg, and Salant 2014). Therefore, decreases in drilling eventually lead to lower production and lower severance tax revenues, which has led to the current dramatic decrease in severance tax revenues in Wyoming, and is further compounded by persistent low gas prices since 2012. The decrease in severance tax revenues has led Wyoming to slash its education budget and may lead to further cuts in the future (Bendix 2017).

In order to provide some counterfactual context to our public-sector analysis we have also included the results of univariate regressions of severance taxes on natural gas prices in the states of Utah and South Dakota. The results of these regressions are presented in the last two columns of table 12.2. The estimated coefficient for Utah is positive and statistically significant, but much smaller in magnitude than the coefficient in Wyoming. The estimated coefficient for South Dakota is actually negative, but statistically insignificant. These estimates highlight the degree to which tax revenues in Wyoming are dependent on natural gas prices relative to nearby states. It would be interesting to repeat our analysis over the pre- and postshale eras as we did in the private-sector analysis; however the annual severance tax data makes this infeasible.

It is interesting to compare and contrast the findings of our private-sector analysis with the findings of our public-sector analysis. We found that natural gas prices are the most important predictor of job creation in Wyoming, and we found that natural gas prices are the most important predictor of severance tax revenues in Wyoming. Natural gas futures prices are widely determined outside the state of Wyoming, so the fact that natural gas prices are the primary indicator of job creation and tax revenues associated with natural gas in the state of Wyoming indicates that changes in job creation and tax revenues associated with natural gas will also widely be determined outside of the state. The primary difference between the private-sector and public-sector analyses is that the impact of production on private-sector job creation disappears when prices are controlled for, but the impact of production on severance tax revenues increases when prices are controlled for. These results are consistent with higher natural gas prices increasing the wealth of Wyoming's citizens through resource ownership, and expanded economic activity, even during periods of relatively low drilling and production. During periods of low prices, Wyoming's residents have the cushion of lower utility bills. However, severance tax revenues are only collected when natural gas producers are producing, so the state's tax revenues are dependent on both prices and production—a precarious situation for Wyoming policy makers given that volatile natural gas prices are widely determined outside Wyoming.

Policy Discussion

In the previous section, we found that job creation and tax revenues in the state of Wyoming both depend on the natural gas industry. The primary natural gas related driver of private-sector job creation in Wyoming is natural gas price, while the primary drivers of tax revenues, unsurprisingly, are natural gas price and natural gas production. While we would expect low natural gas prices that have predominated in recent years to be associated with lower job creation in Wyoming, consumers at least continue to experience the benefit of low prices for home heating, and industrial production capacity is expanding in the state to take advantage of the possibility of continuing low prices in the future. On the other hand, the dependence of the state's budget on both high gas prices and high production is likely to continue to lead to problems for Wyoming without serious reforms. Further, any reversal of the recent expansion of Wyoming's government could compound problems in the private sector associated with low gas prices.

Most analysts and natural gas market observers believe that natural gas prices will remain low for a long period because of the shale gas revolution (Aguilera and Radetzki 2016, 79–80; Braziel 2016, 57–59). Eventually, natural gas production in Wyoming may pick up again as more favorable drilling prospects in other regions are depleted. It is also possible that further technological advancement will result in another boom cycle in Wyoming. However, what is clear is that the boom and bust cycles lead to uncertain state budget situations, and that this budget uncertainty has had a particularly negative effect on Wyoming's education system. The need for diversifying the tax revenue base has long been discussed in Wyoming, but policy makers continue to shy away from diversification, which would require increasing taxes on non-mineral-producing industries (Graham 2018). Further, diversification of the private-sector economy away from dependence on natural gas and other resource extraction requires a strong education system, but Wyoming's education system is heavily dependent on severance taxes collected from resource extraction.

There is currently a push in Wyoming to increase the size of the state's mineral sovereign wealth fund (Western 2016). This would benefit Wyoming by transforming an undiversified portfolio of assets that has long been coal, natural gas, and oil into a more diversified investment portfolio. This approach has been successful in other economies with a large natural resource extraction sector, such as Norway. These include investments that will result in not only more diverse financial instruments but also investments that will lead to a more diverse economy in Wyoming (Western 2016). The primary problem with this approach is that politicians and policy makers tend to turn their focus toward this problem of lack of diversification only during commodity price downturns, while the time for action is generally during booms in commodity prices and production.

In order to avoid reentering the boom and bust cycle that has defined Wyoming's budget in past decades, and to avoid the potential for spillovers from the public sector into the private sector when busts eventually arrive, Wyoming needs to redesign its tax system. In our opinion, the first best solution would involve a tax code that is not dependent on specific industries. The dependence of Wyoming's tax revenues on natural gas, oil, and coal extraction makes the state susceptible to some of the most volative markets in the world, and as a relatively small producer on a global scale, the state of Wyoming cannot enact policies that will dampen volatility in these markets. Wyoming's experience can also inform other states, especially those with newfound resource wealth related to shale gas and tight oil.

Summary of Findings

In this chapter, we analyzed the economic impacts of new natural gas extraction technologies on the state of Wyoming. A brief summary of Wyoming's historical experience with natural gas production from tight reservoirs highlighted the state's unique experience as an early adopter of new technologies, and as a victim of its own medicine when those technologies evolved and unlocked massive quantities of natural gas in other regions. While the private sector in Wyoming has seen job losses as a result of low natural gas prices, consumers continue to benefit from lower utility bills, and commercial and industrial consumers take advantage of low prices by expanding their operations. The public sector, on the other hand, is currently experiencing a financial crisis as a result of depressed natural gas prices and natural gas production in Wyoming. The crisis is due to the state's dependence on severance taxes, and the dependence of severance taxes on natural gas prices and production, as illustrated in our analysis. Additionally, the COVID crisis is likely to exacerbate Wyoming's public-sector budget problems. A recent analysis by the state's Legislative Service Office projects that revenues could fall by $0.5 billion to $2.18 billion by the end of fiscal year 2022 (Angell 2020).

The key policy implication of our results is that the state of Wyoming's dependence on natural gas prices and production put the state's budget in a precarious situation. Natural gas prices are primarily determined outside the state of Wyoming in the much larger US market that is increasingly dominated by the new shale giants such as Pennsylvania. Natural gas production rates in Wyoming are widely determined by the array of economical drilling prospects available in the state given prevailing prices during a given time period. In other words, the two primary determinants of severance tax revenues related to natural gas in Wyoming are out of the control of state policy makers. This precarious budget situation not only engenders instability in Wyoming's fiscal situation but could easily spill over into the private economy by compounding downturns associated with low natural gas prices through sudden budget cuts and uncertainty associated with future government investment. These problems are occurring in Wyoming today. Policy makers in Wyoming should work to broaden the tax base in Wyoming and diversify the state's budget away from commodities as soon as possible in order to alleviate the impact of future boom and bust cycles that are sure to come.

Notes

1. https://www.eia.gov/state/seds/data.php?incfile=/state/seds/sep_sum/html/rank_use_capita.html&sid=US.

2. See https://www.eia.gov/dnav/ng/ng_cons_sum_dcu_SWY_m.htm.

3. Horizontal drilling is technically a subset of directional drilling. Horizontal drilling occurs when the departure of the wellbore from vertical exceeds 80 degrees. See Schlumberger 2018. However, the oil and gas industry generally use the term directional drilling for wells whose wellbore departs from vertical but cannot be categorized as horizontal. Most tight natural gas wells in Wyoming are directional by this definition.

4. Source: https://www.eia.gov/dnav/ng/hist/n9010wy2a.htm.

5. See https://www.eia.gov/dnav/ng/ng_pri_fut_s1_d.htm.

6. See https://wogcc.wyo.gov.

7. We also ran the full regression including oil future prices and the results are qualitatively similar. The main difference is the magnitude of the gas future price is smaller in magnitude, because oil and gas prices are positively correlated. The inclusion of oil future prices did not significantly improve the fit of the model.

8. Severance taxes are applied to the total revenue associated with oil and gas production, so they vary with both prices and production. The severance tax rate in Wyoming is 6 percent (WYDOR 2010).

9. US Census Bureau. 2016.

10. The severance tax rate in Utah varies between 3 percent and 5 percent. The severance tax rate in South Dakota is 4.5 percent (Tax Policy Center 2016).

11. See https://www.census.gov/programs-surveys/stc.html.

References

Aguilera, Roberto F., and Marian Radetzki. 2016. *The Price of Oil*. Cambridge: Cambridge University Press.

Anderson, Soren T., Ryan Kellogg, and Stephen W. Salant. 2014. "Hotelling under Pressure." *Journal of Political Economy* 126 (3): 984–1026. https://doi.org/10.1086/697203.

Angell, Jim. 2020. "Coronavirus Impact: Wyoming Revenues to Drop between $550 Million and $2.8 Billion." *Cowboy State Daily*, April 14. https://cowboys tatedaily.com/2020/04/14/coronavirus-impact-wyoming-revenues-to-drop-between-550-million-3-billion/.

Bendix, Aria. 2017. "Why Oil and Coal States Are Slashing Their Education Budgets." *Atlantic*, March 15. https://www.theatlantic.com/education/archive/2017/03/why-oil-and-coal-states-are-slashing-their-education-budgets/519738/.

Bleizeffer, Dustin. 2010. "Wyoming Approves 'Fracking' Disclosure Rules." *Casper Star Tribune*, June 9. https://trib.com/news/state-and-regional/wyoming-approves-fracking-disclosure-rules/article_069139a4-5b9b-51c3-a599-a38f788e8ff4.html.

Braziel, E. Rusty. 2016. *The Domino Effect*. Houston: NTA Press.

Brown, Jason, Timothy Fitzgerald, and Jeremy Weber. 2016. "Capturing Rents from Natural Resource Abundance: Private Royalties from U.S. Onshore Oil and Gas Production." *Resource and Energy Economics* 46 (November): 23–28. https://doi.org/10.1016/j.reseneeco.2016.07.003.

Brown, Jason, Timothy Fitzgerald, and Jeremy Weber. 2017. "Asset Ownership, Windfalls, and Income: Evidence from Oil and Gas Royalties." Federal Reserve Bank of Kansas City. Research Working Papers. https://doi.org/10.2139/ssrn.2963775.

Davis, Steven J., and John Haltiwanger. 2001. "Sectoral Job Creation and Destruction Responses to Oil Price Changes." *Journal of Monetary Economics* 48 (3): 465–512. https://doi.org/10.1016/S0304-3932(01)00086-1.

Davis, Steven J., John C. Haltiwanger, and Scott Schuh. 1998. *Job Creation and Destruction*. Cambridge, MA: MIT Press.

Gold, Russell. 2014. *The Boom: How Fracking Ignited the American Energy Revolution and Changed the World*. New York: Simon and Schuster.

Graham, Andrew. 2018. "Gubernatorial Candidates Talk Diversification, Not Taxes." WyoFile, June 19. https://www.wyofile.com/gubernatorial-candidates-talk-diversification-not-taxes/.

Kaminski, Vincent. 2012. *Energy Markets*. London: Risk Books.

Kunce, Mitch. 2003. "Effectiveness of Severance Tax Incentives in the U.S. Oil Industry." *International Tax and Public Finance* 10:565–87. https://doi.org/10.1023/A:10261 22323810.

Law, Beverly E., and Charles Worthen Spencer. 1993. "Gas in Tight Reservoirs: An Emerging Major Source of Energy." US Geological Survey. Professional Paper 1570:233–52.

Lockner, Allyn O. 1964. "The Economic Effect of the Severance Tax on Decisions of the Mining Firm." *Natural Resources Journal* 4 (3): 468–85. https://www.jstor.org/stable/24879780?seq=1.

Maniloff, Peter, and Dale T. Manning. 2018. "Jurisdictional Tax Competition and the Division of Nonrenewable Resource Rents." *Environmental and Resource Economics* 71:179–204. https://doi.org/10.1007/s10640-017-0143-6.

Mason, Charles F., and Gavin Roberts. 2018. "Price Elasticity of Supply and Productivity: An Analysis of Natural Gas Wells in Wyoming." *Energy Journal* 39, Special Issue 1 (June): 79–95. https://doi.org/10.5547/01956574.39.SI1.

Noble, Ann C. 2014. "The Jonah Field and Pinedale Anticline: A Natural-Gas Success Story." WyoHistory.org, November 8. https://www.wyohistory.org/encyclopedia/jonah-field-and-pinedale-anticline-natural-gas-success-story.

Schlumberger. 2018. "Horizontal Drilling." Schlumberger Oilfield Glossary. http://www.glossary.oilfield.slb.com/Terms/h/horizontal_drilling.aspx.

Tax Policy Center. 2016. "Briefing Book." Accessed June 7, 2018. https://www.taxpolicycenter.org/briefing-book.

US Bureau of Labor Statistics. 2019. "Business Employment Dynamics." https://www.bls.gov/bdm/.

US Census Bureau. 2016. "Annual Survey of State Government Tax Collections (STC)." Accessed June 3, 2018. https://www.census.gov/programs-surveys/stc.html.

Wen, Sherry. 2018. "Wyoming Labor Force Trends." Wyoming Department of Workforce Services. Accessed June 6, 2018. https://doe.state.wy.us/lmi/trends/0318/0318.pdf.

Western, Samuel. 2016. "Wyoming Considers Changes in Investment Strategy." *Casper Star Tribune*, September 18. https://trib.com/news/state-and-regional/wyoming-considers-changes-in-investment-strategy/article_12a064de-3556-52db-8a07-454bb6f458e6.html.

WSGS. 2018. "Wyoming's Oil and Gas Facts." Wyoming State Geological Survey. Accessed June 3, 2018. http://www.wsgs.wyo.gov/energy/oil-gas-facts.

WYDOR. 2010. "2010 Production Petroleum Severance Tax Reporting." Wyoming Department of Revenue. Accessed May 27, 2018. http://revenue.wyo.gov/mineral-tax-division/severance-tax-filing-information.

13

AN ECONOMIC AND POLICY ANALYSIS OF SHALE GAS WELL BONDS

Max Harleman

Over the last decade, citizens have voiced their concerns about the immediate impacts of shale development, such as those related to air and water contamination during the drilling and hydraulic fracturing processes. In many states, officials have responded by devising new laws and regulations to govern the industry. But relegated to the periphery of the shale debate is the issue of cleaning up wells after the boom subsides. Shale wells do not eternally produce oil or gas—production declines over time, and operators will want to cease to maintain wells when they stop producing profitable quantities. Nor do shale fields produce eternally—researchers estimate that production has already peaked in the nation's first heavily developed field, the Barnett in Texas (Browning et al. 2013).

When wells stop producing oil or gas, most states require operators to "reclaim" wells by plugging them with cement, removing equipment, and restoring surrounding land, water, and vegetation to predrilling conditions. But many wells drilled in past eras have not been reclaimed: they are "abandoned." For wells drilled in the late nineteenth and early twentieth centuries, this is largely because of lenient enforcement of reclamation laws. More recently, states have begun to more strictly enforce these laws by ordering operators to reclaim wells or face civil penalties (PA DEP 2018a; PA DEP 2018b; Sisk 2018). But if the operator has gone out of business or is no longer identifiable, abandoned wells become the responsibility of the state. This is a problem because state funds to reclaim wells are limited, and numerous studies suggest that abandoned wells pose risks to the natural environment and human health. Abandoned wells can leak harmful gases and liquids, provide pathways for the contamination of water at the surface and underground, fragment natural habitats, and interfere with alternative land uses (Ho et al. 2016).

In light of these risks, many states impose bonding requirements to incentivize operators to reclaim wells. Bonding requires operators to set aside funds (or to ensure the state that they have access to funds) before drilling a well, which are forfeited to the state if the well is abandoned. States use forfeited bonds to plug wells, restore land, and remove equipment. The challenges facing state officials when setting bonds for shale gas wells are twofold. First, there is considerable uncertainty surrounding the cost of reclaiming them. Shale gas wells are drilled deeper and utilize larger amounts of land and equipment, which could make them costlier to reclaim than wells from previous eras. The potentially high reclamation costs and environmental risks from abandoned shale gas wells have led stakeholders to consider whether current bonds are large enough to cover cleanup.

Second, officials face a trade-off. On one hand, if bonds are lower than reclamation costs, increasing them could prevent taxpayers from bearing the costs of cleanup. It would also ensure that the state has adequate funds to reclaim wells to prevent damages to human health and the environment. On the other hand, increasing bonds imposes additional costs on operators, which could cause them to drill fewer wells. Less drilling would reduce economic benefits that are received by residents, such as wages paid to oil and gas workers or royalties paid to landowners.

The chapter in this volume by McClure, Simonides, and Weber explores the first of these challenges by providing a better understanding of the cost of reclaiming wells. This chapter explores the second challenge by aiming to answer the question: How much money should state officials require shale gas well operators to set aside in the form of bonds to achieve a socially desirable outcome? As in most cost-benefit analyses conducted by academics and government officials, I consider efficiency, or the maximization of net benefits, as the primary criterion for determining what is socially desirable. In this case, net benefits are defined as payments to citizens relative to the environmental harms they experience. As another criterion, I consider equity, or the distribution of reclamation costs across operators and taxpayers. Bonding requires operators to bear costs that would otherwise be borne by private landowners or taxpayers. Because all states with extensive oil and gas development require operators to reclaim wells and post bonds, I assume that citizens believe that reclamation is socially desirable, and that taxpayers should be shielded from bearing reclamation costs.

Applying these two criteria to the case of bonding and the shale gas industry in Pennsylvania allows us to understand the factors that state officials might consider when selecting a bonding system. The case study approach allows us to see that decisions made by state officials will have direct implications on whether shale gas wells are properly reclaimed, which in turn will affect residents of the state and elsewhere. I construct a simple framework for monetizing the impacts of alternative bonding systems using assumptions based on evidence from the state's history of oil and gas development. Monetizing impacts of a policy means applying a dollar value to both its costs—that is, the forgone wages paid to laborers if a well is not drilled—and its benefits—that is, the value of agricultural land that has been preserved if a leaking well is plugged. Full details on monetization and calculations of net benefits, including underlying data, models, and sensitivity tests, can be found in Harleman (2018). In this chapter, I present the categories of costs and benefits that are relevant for an analysis of alternative bonding systems and present the key results to shed light on how officials can select a socially desirable bonding system.

Pennsylvania as a Case Study for Selecting a Bonding System

Pennsylvania is an appropriate case for studying bonding systems for shale gas wells for three reasons. First, Pennsylvania currently contains between 470,000 and 750,000 wells, and one-third of them have not been properly reclaimed (Kang et al. 2016). We can abstract from Pennsylvania's long history of conventional oil and gas development, which spans more than a century and a half, in order to make assumptions about how likely it is that well operators will abandon shale gas wells, and how likely it is that abandoned wells will pose environmental harm.

Second, more than 10,000 shale gas wells have been drilled in Pennsylvania in the last decade, and projections indicate that more shale gas will be produced there than anywhere else in the country over the next two decades (Staub 2015). Fewer than 1,000 shale gas wells have been plugged in Pennsylvania so far, and only six have been abandoned (Weber and Earle 2019). It is impossible to predict how many shale gas wells will be abandoned going forward. But if bonds are low relative to the cost of reclamation, operators may be incentivized to abandon wells, and forfeited bonds may be insufficient to cover reclamation. The state may also be stuck with a large bill if bonds are too low and many operators become insolvent before reclaiming wells. Although most wells are drilled by highly capitalized companies, research suggests that there is some risk of widespread insolvency because wells are sold to less capitalized owners when production becomes economically unprofitable (Mitchell and Casman 2011).

Third, with the surfeit of abandoned wells from previous eras serving as a salient reminder of the failed policies of the past, officials in Pennsylvania have proposed policies to increase the amount that operators must set aside as bonds for shale wells. Comparing these proposed alternatives to the "status quo" allows us to consider whether state officials can produce a more socially desirable outcome by revising the bonding policy.

Current bonding requirements (the "status quo") are set forth in Pennsylvania Act 13 of 2012. Before issuing a permit for drilling, the Pennsylvania Department of Environmental Protection (DEP) requires well operators to commit bonds. The total bond amount owed by a well operator depends on the number of wells it has in production. Table 13.1 shows the bond amount that an operator would owe for the typical shale gas well. For example, a company that operates thirty wells would pay a "blanket bond" of $140,000 for its first twenty-six wells, and an additional $10,000 for the remaining four wells, for a total amount of $180,000. The act includes caps on total amounts: a company operating up to 150 wells would pay no more than $430,000.

In cases where operators forfeit bonds, they remain legally responsible for violating laws related to reclamation. The DEP may impose penalties or deny permits for new wells if an operator violates provisions related to specific reclamation activities. Total penalties on operators can range from a few thousand dollars to hundreds of thousands. However, if a well operating company declares bankruptcy there may be insufficient assets to pay penalties. In such cases, bonds are meant to fully cover reclamation costs.

Two alternative systems have been proposed to replace the status quo, both of which do away with blanket bonds and caps and increase the amount of money that operators must set aside. The first, a "$400,000" system, has been proposed in the Pennsylvania state legislature. The system would increase bonds to $400,000 for each shale gas well.[1] The second, a "per foot" system, is intended to align

TABLE 13.1   Bonding requirements for shale gas wells in Pennsylvania

BONDING REQUIREMENTS	NUMBER OF WELLS			
	FEWER THAN 26	26 TO 50	51 TO 150	GREATER THAN 150
Blanket bond	-	$140,000	$290,000	$430,000
Amount per additional well	$10,000	$10,000	$10,000	$10,000
Maximum amount	$140,000	$290,000	$430,000	$600,000

Source: Pennsylvania Act 13 of 2012: Oil and Gas Omnibus Amendments. No. 58, Pa. Consolidated Statutes, §3225.

bonds with the factors that influence reclamation costs, an idea that has been supported by researchers and environmentalists (Joyce 2015; Ho et al. 2016). The largest cost associated with reclamation is for plugging, which depends on the vertical depth of the shale formation. At present, the best estimates of the costs to plug shale gas wells in Pennsylvania are those found in the chapter in this volume by McClure, Simonides, and Weber. The authors use data on conventional wells to show that costs depend on the number of wells included in a plugging contract and describe two plausible scenarios where plugging shale gas wells costs between about $15 and $20 per foot. I use $17.5 per foot of vertical depth, the central point of this interval, to define the amount that operators must set aside in a per foot system. I also consider all three systems under a scenario where shale gas wells are much costlier to reclaim.

The three systems represent the wide array of bonds that can be selected by state officials. The status quo system, defined by blanket bonds and caps, is similar to many systems employed by major oil and gas producing states. As we will see, it requires small bonds relative to the anticipated costs of reclamation. The per foot system raises the amount to a level consistent with the best available estimate of well plugging. The $400,000 system raises the bar even higher, in anticipation of shale well reclamation costs that are greater than those for conventional wells.

Four considerations highlight the structure of the analysis that state officials should use to compare alternatives to the status quo. First, it is appropriate for state officials to focus exclusively on marginal impacts. Marginal analysis compares only the benefits and costs driven by a policy change to determine the alternative that maximizes net benefits. For instance, officials should consider the change in the number of abandoned wells under each alternative, and count avoided environmental impacts associated with greater reclamation as economic benefits. The economic costs and benefits associated with the oil and gas industry as a whole will be the same across the alternatives and should be excluded from the analysis. Several other studies consider the net benefits of the shale gas industry. They find positive welfare impacts due primarily to lower prices paid by end-users of natural gas and electricity (Hausman and Kellogg 2015), and present categories of costs that active shale gas wells can pose to human health and the environment (Sovacool 2014; Mason, Muehlenbachs, and Olmstead 2015). In contrast, an analysis of bonding systems should center on the potential impacts that may arise if shale gas wells become abandoned.

Second, officials must decide on a relevant time horizon for the analysis. Should officials consider only costs and benefits experienced by current generations, or should they seek to preserve resources for future generations? The answers to these questions depend on the values held by officials and their constituents. To avoid such moral quandaries, officials might choose to compare systems over a range with available projections about future oil and gas markets. A reasonable approach would be to use the thirty-year time horizon used by the US Energy Information Administration (EIA) in their "Annual Energy Outlook," which is widely regarded as the most comprehensive projection of domestic energy markets (US EIA 2017). Accordingly, this analysis draws on EIA data and uses a horizon which begins in 2020 and ends in 2050.

Third, after selecting a time horizon, officials must decide how to weigh benefits or costs that occur in the present versus those that occur many years in the future. In practical terms, this is a decision about the discount rate to utilize in the analysis. With a low discount rate, benefits and costs that are experienced many years in the future hold more weight. The selection of a discount rate is subjective; if officials and their constituents care more about costs and benefits that occur in the present, they should choose a high rate. Alternatively, if costs and benefits experienced in the future are of higher priority, they should choose a low rate. The results presented below rely on a discount rate of 3 percent, which is consistent with the Interagency Working Group on the Social Cost of Greenhouse Gases (2010). In order to encompass differing opinions on how to value damages that occur far in the future, I also discuss how the results change using discount rates of 2.5 and 5 percent, a range which is inclusive of the state government's rate of 4.5 percent (Pennsylvania Independent Fiscal Office 2014).

Fourth and finally, state officials must decide which people have standing. Should they only count costs and benefits that accrue to state residents, or also those that accrue to people living elsewhere? Some might argue that only Pennsylvania residents should hold sway in state decisions. But given the state's efforts to reduce carbon emissions and methane leaks to abate global warming, it can be assumed that at least some residents and officials view themselves as global citizens, and care about costs that accrue outside the state (PA DEP 2019; Wolf 2016). State officials and citizens may be intrinsically motivated to prevent environmental harms that accrue elsewhere, or they may anticipate that doing so will encourage officials elsewhere to act reciprocally, leading to desirable outcomes for the collective whole (Ostrom 2000). This analysis will primarily be conducted from the state perspective, but I will also discuss how the alternatives perform from a global perspective. Adopting a broader perspective will allow us to determine whether incentives to reclaim wells in Pennsylvania are in harmony or discord with incentives at a higher level.

The Economic Benefits of Increasing Bonds

With the framework set for the case study of Pennsylvania's bonding system, we can begin to consider the benefits that would be experienced if the state increased bonds to the $400,000 or per foot system. Officials can view the benefits of increasing bonds as falling into two categories: avoided reclamation costs borne by taxpayers and avoided environmental damage.

Using data from the PA DEP (2017),[2] I estimate that about $2,000 is posted for the typical well in the state, calculated as the average bond amount per well across all operators and weighted by the number of active wells held by each operator. However, this weighted average is driven down by operators with many wells, who pay relatively small bond amounts due to blanket bonding. It is reasonable to assume that the primary driver of future abandonment will be smaller, less capitalized operators going out of business without transferring their wells to other companies. Calculating the same weighted average for only "small operators" (those in the bottom three quartiles of ownership that own 20 percent of active wells) results in a good estimate of the average bond amount for an abandoned well: about $4,700. This is much less than the average cost of reclaiming a well—about $110,000, given the $17.5 per foot estimate and an average depth of 6,300 for the typical shale well in Pennsylvania as found by McClure, Simonides, and Weber. If a well was abandoned, the difference would be paid by taxpayers. If bonds are increased to match or exceed reclamation costs, the burden of reclaiming wells would be shifted to operators. If many shale gas wells are abandoned, this could save taxpayers millions of dollars. These savings can be viewed as a benefit from both the state and global perspectives, because Pennsylvania taxpayers are simultaneously state and global citizens.

In addition to relieving taxpayers from bearing cleanup costs, increasing bonds can benefit residents by allowing the state to quickly reclaim wells, which can prevent environmental damage. One of the primary environmental concerns with abandoned wells is that they can leak substantial amounts of methane: Kang et al. (2016) estimate that they account for 5 to 8 percent of man-made methane emissions in Pennsylvania. Methane is a potent greenhouse gas that can accelerate climate change, which scientists predict will cause considerable economic, biological, and human costs (Intergovernmental Panel on Climate Change 2014). If increasing bonds provides the state the funds needed to quickly reclaim wells, the amount of time that they leak methane would be reduced. This reduction can be viewed as a benefit from the global perspective, because the impacts of climate change would be dispersed widely across the earth. But from the state perspective, officials might place a relatively small value on emissions reductions, based on the logic that damages in Pennsylvania would make up an extremely small share of global damages.

In addition to methane emissions, abandoned wells can impose environmental costs that are experienced more locally. Unplugged wells can leak brine, oil, and gas that can contaminate soil and water (Dilmore et al. 2015). They can allow gas to migrate to near-surface environments, presenting explosive risks in nearby homes. More generally, shale well pads require around nine acres of land (Johnson et al. 2010) and abandoned pads would tie up land that could be used for more productive purposes. Environmental costs associated with land use and contamination can be viewed as an opportunity cost to landowners. If a well pad is abandoned or land is contaminated, the opportunity cost is the lost value that landowners could have received by using their land for other purposes, such as farming or real estate development. Water contamination can be viewed as a direct cost to residents if they must pay for alternative sources of water for domestic or agricultural uses. If bonds do not cover reclamation costs, owners would experience environmental costs from the time that the well is abandoned until it is reclaimed using taxpayer dollars, which could take several years. Increasing bonds would offset some or all of these costs by enabling the state to quickly reclaim wells.

The Economic Costs of Increasing Bonds

Officials should view the costs of increasing bonds as the decrease in economic benefits that would occur relative to the status quo. The most broadly experienced benefit of the boom in shale gas production has come in the form of reduced domestic natural gas prices. For the typical consumer, this has meant lower electricity and home heating bills. Increasing bonds would impose additional costs on operators, which could cause them to drill fewer wells. This in turn could reduce the total amount of gas produced, which could theoretically raise prices paid by consumers. However, it is reasonable to assume that increasing bonds would have a negligible impact on prices.

Forgone production of gas would represent a small fraction of national supply. In the case of the $400,000 system, the capital necessary to drill the average shale well would increase by $398,000 if an operator staked the entire bond amount as collateral. This is only 4.4 percent of the $9 million in revenues generated by the typical shale gas well in Pennsylvania (Harleman and Weber 2017). It would likely be an even smaller fraction because most operators would not stake collateral and would instead buy a bond from a corporate surety for some percentage of the total bond amount. Newell, Prest, and Vissing (2016) find that the elasticity of natural gas drilling is 0.7—as the price of natural gas decreases by one percent, operators drill 0.7 percent fewer wells. If increasing bonds is thought of as a 4.4 percent reduction in the price received by operators, applying this elasticity suggests that operators would drill 3.1 percent fewer wells. Pennsylvania operators produce 20 percent of the domestic supply of natural gas (US EIA 2020). If 3.1 percent less drilling resulted in 3.1 percent less gas produced in Pennsylvania, domestic supply would decrease by less than one percent. But well productivity varies greatly across the state (Cook and Van Wagener 2014). If faced with increased bonds, operators would forgo drilling in the least productive areas, meaning that 3.1 percent less drilling would result in a less than 3.1 percent decrease in production. Domestic supply could therefore be expected to decrease by less than half a percent, which would have a negligible effect on domestic prices.

Instead of considering impacts on natural gas prices, officials should focus on two categories of economic costs associated with forgone payments to residents: wages and royalties. Wages are payments made by oil and gas companies to their employees. Royalties are payments made by gas companies to mineral owners and are typically calculated as a share of the value of gas that is extracted. From the global perspective, both wages and royalties should be ignored because they are simply transfers between parties. But from the state perspective, wages and royalties are benefits because they are payments to Pennsylvania residents that are financed primarily by consumers of natural gas in other states. If operators decide not to drill wells in response to increased bonds, these payments would be forgone.

Three points outline how officials should view forgone wages. First, forgone wages paid to state residents should be considered in their entirety, while only the share of forgone wages paid to nonresidents that are spent in Pennsylvania should be considered. Brundage et al. (2011) find that 63 percent of the oil and gas workers in Pennsylvania are residents, and Kelsey et al. (2011) find nonresident workers spend 50 percent of their earnings in Pennsylvania, so 81.5 percent of wages paid by operators can be viewed as benefiting Pennsylvania.

Second, only the proportion of forgone wages that represent net social costs should be counted. Bartik (2015) provides a useful framework for considering the social costs associated with job losses created by environmental regulations. If bonding causes wells to be forgone, some job-losers will find a job in another industry. Even if they earn less relative to the wage rate in the natural gas industry, this differential can be ignored because in a competitive labor market it would be offset by the incremental costs they would otherwise incur, such as those associated with higher risks of gas-related employment. It may also be the case that less labor demand will reduce wages across the state, but these changes can be ignored because lower wages are simply transfers from employees to employers. What is more important is that some job-losers will not quickly find another job, and instead become involuntarily unemployed. Involuntary unemployment can be viewed as a net cost, equal to the employees' lost wages plus the value of reductions to their well-being and future earnings potential. Lost tax revenues caused by decreased labor force participation can also be counted as a net cost. Bartik estimates that in an area with the median unemployment rate, the total social cost of permanent job losses ranges from 12.8 percent to 17.5 percent of the change in earnings. I use 15 percent, the central point of this interval, to represent the proportion of lost earnings that represents social costs of job losses associated with increasing bonds.

Third, forgone wages should be valued based on the wages paid at the typical wellsite. Each well in Pennsylvania provides about thirteen full-time jobs over the year that it is drilled (Brundage et al. 2010). The timeline from site construction to completion varies widely but based on a typical drilling timeline I assume that full-time employment for these employees lasts for one year, or 2,087 work hours (US Office of Personnel Management 2017). Dry gas wells require 0.18 full-time employees once they reach the production phase (Brundage et al. 2010). Forgone payments to these employees should be applied over the lifetime of the well, starting in the year it would have been drilled under the status quo policy. I assume that all employees receive Pennsylvania's median hourly wage of about $18 (US Bureau of Labor Statistics 2018).

Reduced drilling could also cause residents to miss out on royalties. Only royalties paid to landowners that are state residents are relevant. Pennsylvania residents own 80 percent of the acreage that has been leased to well operators in the state (Harleman and Weber 2017). If a well is forgone, the mineral owners will be unable to receive royalties in the present. But they still own the natural gas that is beneath the ground. As previously discussed, increased bond amounts can be thought of as a 4.4 percent reduction in the price received by operators for the typical well. It is possible that this reduction could cause some marginal wells to never be drilled. But prices regularly fluctuate (between 2016 and 2017 average annual natural gas spot prices increased by about 19 percent (US EIA 2018)) and it is highly likely that after the study period prices will increase enough that many "previously forgone" wells will become economical, especially as the "low-hanging fruit" in the Marcellus and other natural gas plays has been extracted. If these wells become economical, mineral owners will again be presented with the opportunity to receive royalties. Due to the high likelihood that most natural gas in the ground will have value in the future, it seems reasonable to consider only a proportion of wells that will never be drilled to obtain the value of forgone royalties due to increased bonds. With this in mind, I assume that half of wells forgone in the study period will never be drilled. Although this share may seem high given the ratio of bonds to revenues, it countervails the lower time value of royalties received in the future relative to those received in the present.

Officials might also want to consider the fees and taxes that companies pay to the state for permitting and operating wells. For example, in Pennsylvania operators pay an annual "impact fee" that can contribute over $300,000 per well. However, these fees and taxes are designed to pay for public administrative costs and to offset negative impacts that are related to the industry, such as repairing roads damaged by the industry's use of heavy trucks. If fees and taxes are set at the correct level, they should only cover the social costs that are imposed by drilling. If drilling is forgone, these social costs would not occur, and it would not be necessary for the state to collect the fees and taxes.

Selecting a Bonding System under Uncertain Conditions

State officials can select a bonding system by determining whether an alternative increases net benefits relative to the status quo. An increase in net benefits is an appropriate criterion because it includes all the impacts of increasing bonds, both favorable and deleterious, on all affected parties. However, net benefits cannot be measured with certainty. Future market conditions, responses by the industry to policy changes, and shifts in the ideals held by officials and their constituents might alter conclusions about which alternative is desirable.

To deal with uncertainty, it is useful to first measure net benefits under a reference case that is based on assumptions made with the best available evidence. I subsequently alter these assumptions and see how the results change. To build a reference case, I assume that it will cost $17.5 per foot to reclaim a shale gas well, that a 3 percent discount rate is an appropriate proxy for how officials weigh impacts that occur in the present versus those in the future, and that operators are no more or less likely to abandon wells if bonds are low versus when they are high. These assumptions, along with evidence about the likely impacts of increasing bonds, set the stage for monetizing the relevant categories of costs and benefits.

The results of the reference case are presented in table 13.2, panel A, which depicts the net benefits of the alternative systems relative to the status quo. Detailed information about how I monetized the costs and benefits presented in table 13.2 can be found in Harleman (2018). Using historical data on well abandonment and future projections of drilling in Pennsylvania, I estimate that 193 wells will be abandoned between 2020 and 2050 under the status quo (table 13.3). Because reclamation costs are expected to exceed the bond on each well, only nine wells are reclaimable with bond funds. From the state perspective, the rest of the 184 wells present taxpayers with over $13 million in reclamation costs and local environmental costs associated with unrestored land and water.

TABLE 13.2 Present value (using a 3 percent discount rate) of costs and benefits for two alternative systems (thousands of dollars)

RECLAMATION COSTS:	PANEL A: $110,000 PER WELL				PANEL B: $700,000 PER WELL			
	$400,000 BOND		PER FOOT ($110,000)		$400,000 BOND		PER FOOT ($110,000)	
BONDING SYSTEMS:	STATE	GLOBAL	STATE	GLOBAL	STATE	GLOBAL	STATE	GLOBAL
BENEFITS VERSUS THE STATUS QUO								
Category 1: Avoided reclamation costs								
Avoided reclamation costs borne by taxpayers	$12,584	$12,584	$12,584	$12,584	$ 47,952	$ 47,952	$ 13,688	$ 13,688
Category 2: Avoided environmental costs								
Avoided costs from methane emissions	NC	$ 979	NC	$ 979	NC	$ 524	NC	$ 95
Avoided costs of unrestored land	$ 812	$ 812	$ 812	$ 812	$ 428	$ 428	$ 77	$ 77
Avoided costs from water contamination	$ 122	$ 122	$ 122	$ 122	$ 64	$ 64	$ 11	$ 11
Costs versus the status quo								
Category 1: Employment effects								
Forgone wage income	$ 3,859	NC	$ 996	NC	$ 3,859	NC	$ 996	NC
Category 2: Production revenues								
Forgone royalty payments	$ 18,654	NC	$ 4,814	NC	$ 18,654	NC	$ 4,814	NC
Total net benefits [benefits minus costs]	**$(8,995)**	**$ 14,497**	**$ 7,708**	**$ 14,497**	**$25,931**	**$48,969**	**$ 7,967**	**$ 13,872**

Note: "NC" indicates that the category was not considered under the perspective (state or global) for reasons described in the text.

TABLE 13.3   Projected annual shale gas well drilling, abandonment, and reclamation, 2020 to 2050

	BONDING SYSTEMS		
WELL CATEGORIES	STATUS QUO	$400,000	PER FOOT
Wells drilled	29,008	28,918	28,985
Wells abandoned	193	190	190
Wells reclaimable with bond funds (reclamation costs $110,000 per well)	9	190	190
Wells reclaimable with bond funds (reclamation costs $700,000 per well)	1	109	30

Both the alternative systems set aside enough funds for the state to reclaim abandoned wells immediately, and fully avoid the reclamation and environmental costs experienced in the status quo. Under the $400,000 system, the state overcollects $34 million beyond what it spends on reclamation—a direct transfer from operators to the state. Because bonds greatly exceed the expected costs of reclamation, the $400,000 system unnecessarily reduces operators' incentives to drill wells. With fewer wells drilled, Pennsylvanians miss out on over $22 million in wages and royalties paid by operators, an amount that greatly exceeds the benefits the system creates in the state. In contrast, the per foot system sets bonds to match average reclamation costs and does not result in inefficient overcollection of bonds. As a result, operators do not reduce drilling as much, only about $6 million in wages and royalties are forgone, and the system creates positive net social benefits. This makes the per foot system socially desirable relative to the status quo in the reference case.

Although the per foot system is socially desirable when bonds match reclamation costs, officials might wonder about the accuracy of the $17.5 per foot estimate, which comes from a study of conventional wells in Pennsylvania. What might happen if officials rely on this estimate to set bonds, and costs to reclaim shale gas wells turn out to be much higher? Reclamation costs for shale gas wells might be higher than older wells because they are drilled deeper, utilize larger amounts of land and equipment, and contain horizontal sections that must be sealed off. Very few shale gas wells have been plugged, but Mitchell and Casman (2011) note that an oil and gas company reported that it spent $700,000 per well to reclaim three active shale gas wells in Pennsylvania. This amount may be artificially high because the DEP ordered the company to reclaim active wells that were found to have contaminated nearby drinking water, and because the company may have overstated reclamation costs to show that it had gone to great lengths to comply. Nevertheless, the $700,000 figure serves as a high-end estimate, which can be used to shed light on a desirable bonding system if costs to plug shale gas wells turn out to be much higher than anticipated.

With this in mind, table 13.2, panel B presents net benefits created by the alternative systems if the average shale gas well costs $700,000 to reclaim. Under both alternatives, bonds would not be large enough to reclaim all of the 190 abandoned wells (table 13.3). But under the $400,000 system, bonds can cover the reclamation of 109 wells, which benefits state residents by offsetting $48 million in reclamation and environmental costs. These benefits greatly outweigh the wages and royalties that are forgone when operators respond with less drilling. In fact, the net social benefits from the $400,000 system outweigh those created by the per foot ($110,000) system. From the state perspective, this makes the $400,000 system the socially desirable choice when reclamation costs are high. From the global perspective, the $400,000 system is also desirable, as it offsets over $0.5 million in climate-related damages from methane emissions, an amount that would be even larger if the social costs of climate change turn out to be very severe.

The success of the $400,000 system when reclamation costs are high suggests that matching bonds to the true costs of reclamation is socially desirable. But in the short term, officials will not know the true costs of reclamation, and the best estimate is represented by the per foot system. Setting bonds higher than actual reclamation costs could result in inefficient overcollection of bonds. To deal with this uncertainty, officials could monitor cost data to get a better idea of the true costs of reclamation as more shale gas wells reach the end of their productive lives and are plugged. The state could implement a flexible system that starts with the $17.5 per foot bond and allows officials to revise bonds to match the cost of reclaiming a typical well. The status quo system actually does allow officials to adjust bonds every two years to reflect projected plugging costs, but they have never done so. Repeating the monetization exercise several times reveals that even if average reclamation costs are higher than $700,000, increasing bonds to match reclamation costs would increase net benefits both in the state and globally. But because forgone wages and royalties are very large when bonds are set this high, further economic modeling would be necessary to determine their optimal level. However, it is unlikely that costs will be this high in the long run, because operators and well plugging contractors would be incentivized to develop more cost-effective reclamation technologies.

Another point of uncertainty is how operators will respond to increased bonds. Officials might presume that if bonds are high relative to reclamation costs, operators might be incentivized to reclaim wells and recoup their bonds. The reference case assumes that operators will abandon 0.8 percent of shale gas wells, which is the abandonment rate of wells drilled between 1985 and 1999.

During the 1990s, the abandonment rate dropped to a decade low of 0.3 percent. I apply this low abandonment rate to the $400,000 system to reflect operators' incentives to reclaim wells when bonds are high. Similarly, I apply a 0.5 percent abandonment rate to the per foot system and keep the status quo at 0.8 percent. Repeating the monetization exercise with these abandonment rates decreases the number of wells that are abandoned in the alternative systems, but the per foot system remains desirable when reclamation costs are $110,000 per well, and the $400,000 system remains desirable when costs are $700,000 per well. The abandonment rate might be much higher if widespread insolvency occurs because many wells are transferred to less capitalized operators (Mitchell and Casman 2011), or if low oil and gas prices, such as those triggered by the 2020 economic crisis, are sustained and force operators out of business. Such scenarios would uniformly increase the benefits associated with avoided reclamation costs borne by taxpayers across the two alternatives, and our conclusions would remain the same.

A final point of uncertainty relates to how state officials and their constituents weigh costs and benefits over time. The reference case uses a discount rate of 3 percent, but this may not adequately reflect how officials weigh benefits or damages that occur far in the future against those that occur in the present. Upon repeating the monetization exercise using discount rates of 2.5 and 5 percent, the per foot system remains desirable when costs are $110,000 per well, and the $400,000 system remains desirable when costs are $700,000 per well.

The results of the cost-benefit analysis suggest that by setting bonds to closely match the anticipated costs of reclamation, Pennsylvania would have to give up relatively little to prevent taxpayers from bearing reclamation and environmental costs, as well as hundreds of thousands of dollars in global damages from methane emissions. Many Pennsylvanians would find this to be equitable and socially desirable.

A Socially Desirable Bonding System

The prospect of industrial developments that exhibit costs at the end of their productive lives, such as a factory that contaminates underlying soil, can pose challenges to state officials. They must balance economic benefits for their constituents against the public bearing the costs of cleaning up waste and abandoned infrastructure many years down the line. In the case of shale gas development, bonding systems are a policy tool that can help officials strike such a balance. If set at the right level, bonds force operators to factor the full cost of reclamation into their decisions to drill wells. This should incentivize most operators to reclaim wells themselves in order to recoup their bonds and avoid penalties associated with noncompliance with state laws. Some operators may become financially unable to reclaim wells, but bonds that match reclamation costs ensure that the state has adequate funds to conduct reclamation.

The issue of bonding is especially relevant in Pennsylvania, as many shale gas wells will be drilled there in the years to come. If state officials set bonds at the right level, they can facilitate the development of a lucrative industry, while preventing the widespread abandonment of wells. The case study of bonding systems in Pennsylvania indicates that current bonds are too low. Increasing them to match the best estimates of reclamation costs is likely to create a socially desirable outcome by preventing taxpayers from bearing reclamation costs, averting local environmental damages, and reducing methane emissions that contribute to global warming. In the majority of plausible scenarios, these benefits outweigh the wages and royalties paid to state residents that would be forgone if operators respond to increased bonds by drilling fewer wells.

It is possible that current estimates of shale gas well reclamation costs are too low because they rely on data about the costs of reclaiming older, conventional wells. But setting bonds artificially high, under the premise that reclaiming shale gas wells will be very costly, could result in inefficient overcollection of bond funds. This needlessly reduces operators' incentives to drill wells and create economic benefits in the state. Going forward officials should closely monitor abandoned shale gas wells and the costs incurred to reclaim them. They should adopt a flexible bonding system that allows bonds to be revised to match reclamation costs.

Under most plausible scenarios, keeping bonds closely tied to the true cost of reclamation results in socially desirable and equitable outcomes. But in the eleven major oil and gas producing states, the typical bond amount is tens of thousands of dollars lower than any reasonable projection of shale well reclamation costs (Lee 2019). What explains the divergence? Qualitative evidence—especially field interviews with key stakeholders and decision makers—would be especially useful to answer this question. The evidence presented in this chapter suggests that the answer has to do with the myriad regulatory challenges that state officials faced during the rapid onset of the shale boom. Affected citizens expressed immediate concerns related to water contamination, air quality, and seismicity. In turn, states responded by updating laws and regulations to account for the intricacies of shale production. Much less attention during those early years was paid to concerns that would manifest far in the future, such as those related to well abandonment. But today, there is increased attention by academia and the media—especially to the issue of methane leakage from abandoned infrastructure—along with stricter state enforcement of reclamation laws. These trends suggest that public awareness will continue to grow, which could lead to increased pressure on state officials to adopt more efficient bonding systems.

Notes

1. In March 2017, Thomas Murt, a member of the Pennsylvania House of Representatives representing the 152nd legislative district, along with five cosponsors, introduced House Bill 943 of 2017. The bill proposes that bond amounts be increased to a fixed amount of $400,000 for each shale gas well, with the large amount selected arbitrarily to spark legislative discussions aimed at setting bonds closer to reclamation costs. The Marcellus Shale Coalition, an industry group, stated that a similar proposal in 2016 was "misguided," and suggested that the status quo requirements were adequate to ensure environmental remediation (Legere 2016). The bill was referred to the Environmental Resources and Energy Committee for discussion on March 23, 2017, where it failed to receive further legislative attention. To date, the issue of well bonding has not received further formal legislative attention.

2. Please contact the author for access to the data used in this portion of the analysis.

References

Bartik, Timothy J. 2015. "The Social Value of Job Loss and Its Effect on the Costs of US Environmental Regulations." *Review of Environmental Economics and Policy* 9 (2): 179–97. https://doi.org/10.1093/reep/rev002.

Browning, John, Svetlana Ikonnikova, Gürcan Gülen, and Scott Tinker. 2013. "Barnett Shale Production Outlook." Society of Petroleum Engineers Economics and Management 5 (3): 89–104. https://doi.org/10.2118/165585-PA.

Brundage, Tracy L., Jeffrey Jacquet, Timothy W. Kelsey, James R. Ladlee, Jeffrey F. Lorson, Larry L. Michael, and Thomas B. Murphy. 2010. "Southwest Pennsylvania Marcellus Shale Workforce Needs Assessment." MSETC Needs Assessment Series. Marcellus Shale Education and Training Center, June. https://www.pct.edu/files/imported/business/shaletec/docs/NeedsAssessmentwithcoverSW.pdf.

Brundage, Tracy L., Timothy W. Kelsey, Janice Lobdell, Larry L. Michael, Jeffrey Jacquet, James R. Ladlee, Jeffrey F. Lorson, and Thomas B. Murphy. 2011. "Pennsylvania Statewide Marcellus Shale Workforce Needs Assessment." MSETC Economic Impact Study. Marcellus Shale Education and Training Center, June. http://pasbdc.org/uploads/media_items/pennsylvania-statewide-marcellus-shale-workforce-needs-assesment-june-2011.original.pdf.

Cook, Troy, and Dana Van Wagener. 2014. "Improving Well Productivity Based Modeling with the Incorporation of Geologic Dependencies." Working Paper Series. US Energy Information Administration. Independent Statistics and Analysis, October 14. http://www.eia.gov/workingpapers/pdf/geologic_dependencies.pdf.

Dilmore, Robert M., James I. Sams III, Deborah Glosser, Kristin M. Carter, and Daniel J. Bain. 2015. "Spatial and Temporal Characteristics of Historical Oil and Gas Wells in Pennsylvania: Implications for New Shale Gas Resources." *Environmental Science and Technology* 49 (20): 12015–23. https://doi.org/10.1021/acs.est.5b00820.

Harleman, Max. 2018. "A Cost Benefit Analysis of Shale Gas Well Bonding Systems in Pennsylvania." USAEE Working Paper Series no. 18-359. United States Association for Energy Economics, October 8. https://doi.org/10.2139/ssrn.3257597.

Harleman, Max, and Jeremy G. Weber. 2017. "Natural Resource Ownership, Financial Gains, and Governance: The Case of Unconventional Gas Development in the UK and the US." *Energy Policy* 111:281–96. https://doi.org/10.1016/j.enpol.2017.09.036.

Hausman, Catherine, and Ryan Kellogg. 2015. "Welfare and Distributional Implications of Shale Gas." *Brookings Papers on Economic Activity*. Brookings Institution. https://www.brookings.edu/wp-content/uploads/2015/03/HausmanText.pdf.

Ho, Jacqueline, Alan Krupnick, Katrina McLaughlin, Clayton Munnings, and Jhih-Shyang Shih. 2016. "Plugging the Gaps in Inactive Well Policy." RFF Reports. Resources for the Future, May 18. http://www.rff.org/research/publications/plugging-gaps-inactive-well-policy.

Interagency Working Group on the Social Cost of Greenhouse Gases. 2010. "Social Cost of Carbon for Regulatory Impact Analysis under Executive Order 12866." Technical Support Document. February. https://www.epa.gov/sites/production/files/2016-12/documents/scc_tsd_2010.pdf.

Intergovernmental Panel on Climate Change. 2014. "Climate Change 2014: Synthesis Report; Summary for Policy Makers." Assessment Report, November 2. https://www.ipcc.ch/pdf/assessment-report/ar5/syr/AR5_SYR_FINAL_SPM.pdf.

Johnson, Nels, Tamara Gagnolet, Rachel Ralls, Ephraim Zimmerman, Brad Eichelberger, Chris Tracey, Ginny Kreitler, Stephanie Orndorff, Jim Tomlinson, Scott Bearer, and Sarah Sargent. 2010. "Pennsylvania Energy Impacts Assessment Report 1: Marcellus Shale Natural Gas and Wind." The Nature Conservancy Pennsylvania Chapter, November 15. https://www.nature.org/media/pa/tnc_energy_analysis.pdf.

Joyce, Stephanie. 2015. "New Bonding Regs for Oil and Gas Leave Environmentalists Unimpressed." *Inside Energy*, December 8. http://insideenergy.org/2015/12/08/new-bonding-regs-for-oil-and-gas-leave-environmentalists-unimpressed/.

Kang, Mary, Shanna Christian, Michael A. Celia, Denise L. Mauzerall, Markus Bill, Alana R. Miller, Yuheng Chen, Mark E. Conrad, Thomas H. Darrah, and Robert B. Jackson. 2016. "Identification and Characterization of High Methane-Emitting Abandoned Oil and Gas Wells." *Proceedings of the National Academy of Sciences* 113 (48): 13636–41. https://doi.org/10.1073/pnas.1605913113.

Kelsey, Timothy W., Martin Shields, James R. Ladlee, and Melissa Ward. 2011. "Economic Impacts of Marcellus Shale in Pennsylvania: Employment and Income in 2009." Marcellus Shale Education and Training Center, August. https://aese.psu.edu/research/centers/cecd/archives/marcellus/economic-impacts-of-marcellus-shale-in-pennsylvania-employment-and-income-in-2009/view.

Lee, Jongseon. 2019. "Bonding Requirements for Oil and Gas Wells: How Does Pennsylvania Compare?" *GSPIA Energy and Environment Blog*, February 28. https://medium.com/@GSPIAe_eBlog/bonding-requirements-for-oil-and-gas-wells-how-does-pennsylvania-compare-7eda5c06500f.

Legere, Laura. 2016. "Bill Would Raise Shale Well Site Bonds." *Pittsburgh Post-Gazette*, August 30. https://www.post-gazette.com/business/powersource/2016/08/30/Bill-would-raise-oil-gas-well-site-bonds-Pennsylvania-Marcellus-shale/stories/201608300010.

Mason, Charles F., Lucija A. Muehlenbachs, and Sheila M. Olmstead. 2015. "The Economics of Shale Gas Development." *Annual Review of Resource Economics* 7 (1): 269–89. https://doi.org/10.1146/annurev-resource-100814-125023.

Mitchell, Austin L., and Elizabeth A. Casman. 2011. "Economic Incentives and Regulatory Framework for Shale Gas Well Site Reclamation in Pennsylvania." *Environmental Science and Technology* 45 (22): 9506–14. https://doi.org/10.1021/es2021796.

Newell, Richard G., Brian C. Prest, and Ashley Vissing. 2016. "Trophy Hunting vs. Manufacturing Energy: The Price-Responsiveness of Shale Gas." NBER Working Paper Series. National Bureau of Economic Research, August. https://doi.org/10.3386/w22532.

Ostrom, Elinor. 2000. "Collective Action and the Evolution of Social Norms." *Journal of Economic Perspectives* 14 (3): 137–58. https://doi.org/10.1257/jep.14.3.137.

Pennsylvania Act 13 of 2012: Oil and Gas Omnibus Amendments. No. 58, PA. C.S. https://www.legis.state.pa.us/WU01/LI/LI/US/HTM/2012/0/0013.HTM.

Pennsylvania Department of Environmental Protection (PA DEP). 2017. "DEP Office of Oil and Gas Management: Spud Report." Accessed November 10. http://cedatareporting.pa.gov/Reportserver/Pages/ReportViewer.aspx?/Public/DEP/OG/SSRS/Spud_External_Data.

PA DEP. 2018a. "Order, in the Matter of CNX Gas Company LLC." Last modified July 20. http://files.dep.state.pa.us/RegionalResources/SWRO/SWROPortalFiles/PlugOrders2018/Signed%20CNX%20Plug%20Order%2007-20-2018.pdf.

PA DEP. 2018b. "Order, in the Matter of XTO Energy Inc." Last modified July 20. http://files.dep.state.pa.us/RegionalResources/SWRO/SWROPortalFiles/PlugOrders2018/Signed%20XTO%20Plug%20Order%2007-20-2018.pdf.

PA DEP. 2019. "Pennsylvania Climate Action Plan." April 29. https://www.dep.pa.gov/Citizens/climate/Pages/PA-Climate-Action-Plan.aspx.

Pennsylvania House Bill No. 943 of 2017. http://www.legis.state.pa.us/cfdocs/billinfo/billinfo.cfm?syear=2017&sind=0&body=H&type=B&bn=0943.

Pennsylvania Independent Fiscal Office. 2014. "Natural Gas Extraction: An Interstate Tax Comparison." Special Reports, March. https://old.taxadmin.org/fta/severance/papers/pa_gastaxstudy_0314.pdf.

Sisk, Amy. 2018. "State Orders Companies to Plug More Than 1,000 Abandoned Oil, Gas Wells." StateImpact, July 30. https://stateimpact.npr.org/pennsylvania/2018/07/25/state-orders-companies-to-plug-1000-abandoned-oil-gas-wells/.

Sovacool, Benjamin K. 2014. "Cornucopia or Curse? Reviewing the Costs and Benefits of Shale Gas Hydraulic Fracturing (Fracking)." *Renewable and Sustainable Energy Reviews* 37:249–64. https://doi.org/10.1016/j.rser.2014.04.068.

Staub, John. 2015. "The Growth of U.S. Natural Gas: An Uncertain Outlook for U.S. and World Supply." Paper presented at the EIA Energy Conference, Washington, DC, June 15.

US Bureau of Labor Statistics. 2018. "May 2017 State Occupational Employment and Wage Estimates." Accessed August 22. https://www.bls.gov/oes/2017/may/oes_pa.htm.

US EIA. 2017. "Annual Energy Outlook." Independent Statistics and Analysis, January 5. https://www.eia.gov/outlooks/archive/aeo17/.

US EIA. 2018. "Energy Information Administration: Henry Hub Natural Gas Spot Price." Accessed August 22. https://www.eia.gov/dnav/ng/hist/rngwhhdA.htm.

US EIA. 2020. "Which States Consume and Produce the Most Natural Gas?" Last modified October 19. https://www.eia.gov/tools/faqs/faq.php?id=46&t=8.

US Office of Personnel Management. 2017. "Fact Sheet: Computing Hourly Rates of Pay Using the 2,087-Hour Divisor." Accessed November 3. https://www.opm.gov/policy-data-oversight/pay-leave/pay-administration/fact-sheets/computing-hourly-rates-of-pay-using-the-2087-hour-divisor/.

Weber, Jeremy G., and Andrew Earle. 2019. "Abandoned Unconventional Natural Gas Wells: A Looming Problem for Pennsylvania?" *GSPIA Energy and Environment Blog*, February 27. https://medium.com/@GSPIAe_eBlog/abandoned-unconventional-natural-gas-wells-a-looming-problem-for-pennsylvania-ce8243aaf8a4.

Wolf, Thomas W. 2016. "Governor Wolf Announces New Methane Rules to Improve Air Quality, Reduce Industry Loss." Last modified January 19. https://www.governor.pa.gov/governor-wolf-announces-new-methane-rules-to-improve-air-quality-reduce-industry-loss/.

LESSONS AND EXTENSIONS

Sabina E. Deitrick and Ilia Murtazashvili

The rise of shale gas production in the United States has been called the "shale era," "shale boom," and "shale revolution." All are apt: each captures the fact that the technique of hydraulic fracturing transformed the United States from a minor player to a world leader in natural gas production. The revolution has also been widely felt in places like Pennsylvania and Texas, the leading shale-producing states, and to a lesser degree in Colorado, Oklahoma, West Virginia, and Wyoming, among other states.

The shale revolution likely caught many by surprise. Fracking has been used since the 1950s to extract shale oil, though "fracking" was not part of the vernacular until the early years of this century and even then, much of the fight over shale gas development did not pick up until gas companies began to frack the Marcellus Shale. Since the boom began, there has been an outpouring of attention to the shale boom by policy makers, planners, activists, academics, and the news media.

Yet the study of the shale boom is in many ways just at its beginning. For a comparison, economists are still debating what ended the great depression over a half century after the Stock Market Crash of 1929 (Romer 1990, 1992). Booms and busts are complicated, including energy booms. Economists now have five decades of data on oil and gas development to determine whether the conventional energy sectors are subject to the "Dutch Disease," whereby the booming sector produces a decline in another, such as in the manufacturing sector, discussed in chapter 9 (Allcott and Keniston 2017). The shale era is young by comparison.

Stepping outside the usual range of outcomes considered in economic studies of boom and bust dynamics—typically employment and earnings—heightens the challenge of understanding a resource boom (Mason, Muehlenbachs, and Olmstead 2015). Such a broader approach is essential to understand the governance, planning, and economic impacts of hydraulic fracturing. There are also practical questions in understanding how local governments ought to respond to the boom once it commences, as well as the plan for a bust, should it occur.

This volume embraced the complexity of the shale boom. It showed that governance systems to some extent have adapted to new challenges and that the institutional and regulatory framework is capable enough to encourage fracking, but also attempting to manage the economic and environmental harms from fracking. Judges have also been somewhat successful in clarifying boundaries of authority between the state and local governments, thereby facilitating more accountable governance of the shale boom. Planners have in many instances dealt with land use challenges under complicated conditions. Though the challenges sometimes invoke the image of Sisyphus pushing the boulder up the hill, planners are far from powerless and in many cases have been able to effectively manage the boom and to make a set of reasonable plans for the potential bust. Our economic studies also show that understanding the economic consequences requires consideration of fracking growth alongside the full range of external effects.

We conclude this volume by first, analyzing the practical implications of the essays of this volume. The shale boom is the subject of much policy debate. It is therefore important for us to highlight for policy makers, planners, and the public interested in shale gas development what seems to work, what has not worked, and where much more is needed in terms of improvements in governance or planning.

The second part of this conclusion considers future directions for analyzing the consequences of the shale boom. One of the features of the shale boom is that there are a whole range of associated outcomes. In this conclusion, we touch on a few of these areas, including effects of the shale boom on social capital, poverty, and health. We also require more knowledge of the effects of global competition on shale boom and bust dynamics in the United States, the role of institutions in explaining who wins and who loses in the race for shale gas, and the politics of fracking.

Lessons

The chapters in this volume offer a range of perspectives that let us discern some lessons and implications from these studies. One of the themes is that while communities are under stress, local governance remains an essential tool to manage the consequences of the shale boom and bust. While there is certainly evidence that shale gas presents novel regulatory challenges, there are institutions in place that, when modified through a deliberative process, enable communities to realize benefits from shale gas development and to some extent avoid its costs. We have also seen the important role of legal change to address conflicts arising with the shale boom, as well as to address challenges arising from social equity issues that are a recurrent theme in many communities where fracking commenced.

The lessons from these studies of governance can be summarized as the importance of adaptation of institutions and policies to address novel challenges arising with the shale boom, but also the essential role of an inclusive process of governance, especially at the local level. Alongside inclusivity, autonomy is necessary for local political decision makers to address challenges and for planners to experiment with solutions and to work as best they can to address economic and social costs that arise from the boom. Lawyers and judges must have a role in the process of legal change that helps to alleviate conflicts over shale gas.

It is also clear that in some contexts, communities are under stress from a lack of resources or expertise and that many challenges extend beyond the confines of any single community or even multimunicipal collaboration. This demand for coordination provides a role for the state and federal governments to regulate shale gas development. Yet as these essays show, state and federal governance does not eliminate the role for local governance in the shale boom. Local planners, lawyers, judges, and municipal officials confront imperfect information and often have the wrong kinds of institutional incentives, but they are also the core of self-governance and have essential roles in any community where fracking is underway.

Another lesson of the shale boom is that it is important to bring more groups to the table. Some of the processes governing fracking were supposed to be inclusive but fell short in practice. This often meant that there was a political process that provided for input through voting, but there were also many public processes in planning and governance that brought more local control. Yet in some instances, representation in the process focused only on voice at the ballot box.

Democracy is an obviously important aspect of collective decision-making, especially when people can vote with their feet to signal to politicians their preferences (Somin 2020). The challenge arises when voters do not have information to make informed choices, in which case the democratic process cannot be relied upon to choose the best policies from the perspective of society (Brennan 2017; Caplan 2006). There is also much work to be done to implement and design policies after elections. Accordingly, one of the lessons is the need to bring people to the policy table, as the case studies have shown, to address more fully the challenges with fracking. It is also clear that in some cases, groups seeking bans on fracking have been unsuccessful. In other contexts, as noted in several chapters, bans were attempted through multiple means through ordinances, charter amendment, and, in the case of New York State, a statewide ban on fracking.

Since fracking is likely to continue in states that already have substantial production, the policy-making considerations ought to bring together policy experts with expertise in fracking and its impacts in planning, economic development, public health, environment, geology, seismology, and engineering for discussions. The perception is that the bargaining has mainly been between gas companies and politicians, without sufficient participation from experts, which is an important issue to address—one that is an ongoing consideration.

Of course, the mere presence of experts does not guarantee success (Koppl 2018). The possibility of expert failure further suggests the importance of local policy experiments and shared knowledge and learning. Tying the ideas together, the best strategy may be polycentric experimentation with regulation, but with continued effort to bring in more cooks to stir the policy-making broth. That way, when policies fail, it will not be for a lack of inclusion of experts.

The economic studies in this volume have several lessons. One is that on balance, the shale boom appears to generate earnings and employment gains and positive externalities in supplier industries. At the same time, there are costs and other negative externalities that must be addressed beyond the much-studied problem of pollution related to fracking, such as what to do with abandoned wells. Of course, the economic analysis of monetized net benefits is never enough to determine what policies societies ought to pursue (Weimer and Vining 2017). For example, any analysis of what is efficient with respect to fracking says little about the distribution of wealth, and cost-benefits studies may not consider all the relevant economic externalities from economic growth—especially because some externalities may not be well understood (Mason, Muehlenbachs, and Olmstead 2015).

Extensions of the Research

One of the features of this volume was a desire to delve more deeply into the experience of communities confronting the shale boom. This allowed us to offer a richer understanding of shale gas development. It also required us to leave several areas to future or additional research.

Social Capital and Fracking

One such area is to consider more extensively the social impacts of fracking, such as the impact on social capital and crime. Social capital refers to the bridging and bonding capital that ties people together (Granovetter 1977). Social capital can be thought of as a part of human capital that makes communities better off (Putnam 2000). It is therefore important to study more systematically how fracking influences the social capital of a community.

The case of Williston, North Dakota (see chapter 6), through the analysis of interactions between social capital and the fracking boom, illustrates the importance of considering these questions. At one time, resource extraction often undermined community social capital, but through good planning and local and state regulations, active rather than reactive strategies strengthened community and social capital. Case studies such as this show how communities can strengthen trust and social capital during the boom and bust of fracking.

Recent work on bottom-up perspectives on crisis response and resiliency emphasizes that successful recovery from disasters, both arising from the fickleness of nature and the folly of man, has a political, economic, and social dimension (Boettke et al. 2007; Sobel and Leeson 2006). These studies find that social capital may explain the extent to which communities are able to respond to crises (Storr, Grube, and Haeffele-Balch 2017; Chamlee-Wright and Storr 2009a; Chamlee-Wright and Storr 2009b). As the timeline of the current boom-bust cycle from recent fracking in shale gas and oil areas extends, we will have opportunities to compare community response and resiliency to shale gas development. Such studies provide an important point of departure to develop additional measures of social capital, as well as for additional fieldwork to understand community response to the shale boom.

Another category of consequences is crime. One expected consequence of shale gas development is an increase in crime. Part of the reason is that more men move to areas with fracking and they are more likely to commit crimes. There is also work which considers the impact of shale gas development on prostitution (Cunningham, DeAngelo, and Smith, forthcoming). Crime is also related to social capital. An open question is whether communities with more social capital confront lower increases in crime with shale booms, as well as how crime may influence a community's social capital.

Beyond Growth and Employment: Fracking, Poverty, and Inequality

Much of the economic analysis of fracking considers its consequences for economic growth or employment. Fewer studies consider how the shale boom affects poverty and inequality. This is an important oversight, as capitalism and inequality are currently gaining more attention in politics and popular debate. The way that the economist Thomas Piketty's (2014) *Capital in the Twenty-First Century* resonated provides some insight into the importance of these ideas. Wealth inequality is an increasing concern, especially wealth concentration among the top 1 percent of the income distribution, but also inequality more generally (Saez and Zucman 2016).

These suggest that an important question is what impacts are there on poverty and inequality in a community. Does fracking lead to increases in county-level poverty rates? What about income inequality? Although there is solid evidence that there is substantial royalty wealth going to some in communities with shale gas development, it remains an open question to consider how wealth is distributed. There have been gains in some equity issues through addressing environmental injustice concerns. As discussed in chapter 3, though poorer counties have economic incentives to develop shale extraction, the benefits are not shared by individual landowners and nearby neighbors, who bear the costs.

Does Fracking Make Us Sicker?

It seems clear enough that the long history of private property rights to minerals in the United States contributed to the shale boom. These rights gave mineral rights owners strong incentives to contract with gas companies, as well as to lobby for public policies supportive of fracking. Economic growth increased as a result. But it is also possible that economic liberties may leave us both richer and sicker (Troesken 2015). Fracking might be an example. Property rights resulted in a revolution in gas production, but we are still sorting out the health consequences. Will all of those liberties come back to haunt us?

There is still much research necessary on this front. The economic and epidemiological studies are finding that fracking reduces birth weight, among other health impacts (Gibbons 2017). These studies, by focusing on the impact of fracking on pregnant women, can make a compelling causal case that fracking leads to public health costs.

These consequences are challenging to fit into the conventional models of boom and bust. Indeed, the studies of the economic consequences do not often include economic externalities, let along the health consequences (Mason, Muehlenbachs, and Olmstead 2015). One implication is that the studies of fracking growth almost certainly overstate the net benefits of these economic activities. There are also dynamic aspects of fracking, such as whether the opportunity for income undermines incentives to invest in human capital—thereby reducing long-run prospects for economic growth in regions with higher levels of shale gas production (Rickman, Wang, and Winters 2017).

On the positive side, the EPA has been active in regulating air pollution from wells. These regulations, as they are implemented, are likely to alleviate some of these challenges. In addition, the health studies mentioned above show that the effects are confined to women living close to wells within one kilometer or so. That suggests a regulatory solution, which would extend setbacks from wells. Though potentially costly, this would also address in a direct way a harm associated with fracking. The cause for optimism is that we know what we ought to regulate, though the pessimism arises from whether we expect the political process to do something with this knowledge.

A much more challenging problem may be that the adverse health effects may occur far from the shale wells. Indeed, recent research suggests that the primary adverse health consequences are found in cities—far from any active wellheads—that suffer lower air quality from fracking (Mayfield et al. 2019). An implication is the importance of research to consider the regional consequences of fracking, as well as regional governance and planning to address these impacts.

Global Production

Our focus has been on the United States in this volume. Fracking, of course, is a global phenomenon. While the United States has charged ahead in the shale revolution, many other countries have seen far less production. The United Kingdom had a moratorium on fracking from 2011 to 2012 but has had (essentially) no production as of 2020. In late 2019, the UK government again set up a moratorium on fracking, following seismic activity at the only active fracking site in the nation. Poland has almost no economically recoverable shale, even though the early estimates suggested it had massive shale reserves. Argentina has had drilling, but nowhere near the rate of the United States. China has very strong incentives to diversify energy production away from coal, but as of this writing, shale gas production has been much lower than anticipated.

All of this means that the United States is still a global leader in shale gas production. It also raises important questions about what would happen if fracking does increase in the rest of the world. In US markets, less drilling would occur, as the market value declines. And, even without that market expansion globally, this has already happened in shale plays across the country, as recent years of glut in the industry have reduced commodity prices, shuttered drilling sites, and left a bust.

The global economic impacts of fracking also raise the possibility of international coordination to regulate fracking. The case for supranational regulation of fracking is clearest when it comes to fugitive methane, which is a truly global problem. It need not be regulated just by the national governments but coordinated among all shale-producing nations. Yet we also know that even with global commons challenges, it is important to consider the possibility of polycentric regulation to combat environmental challenges. Elinor Ostrom (2010) emphasized that global climate change issues ought to be addressed at multiple scales, including at the local level. There are also important design challenges with coming up with appropriate international regulations even when there is agreement on the problem (Aldy, Barrett, and Stavins 2003). Thus, in considering the international dimensions of shale gas, it is important to continue to provide a space for local and medium scale governance entities to participate in the process of designing rules to address global commons challenges.

Wrapping Up

The early fear with fracking was that communities were rushing into a disaster. By most accounts, the early fears have not been realized nor has fracking generated the massive economic development benefits promised by politicians. There still remain challenges, and the jury is still out regarding the benefits and costs of shale gas production. It will be years or even decades before we can assess all relevant benefit and cost categories of shale gas development.

Despite such inevitable uncertainties that arise in analyzing the shale boom, there has been a more robust and inclusive response to challenges posed by shale gas extraction. Communities have in general been able to decide on their collective futures through deliberative processes and shared expertise. To be sure, some communities had little choice but to accept fracking, yet others have been able to prevent it. Even in those areas where communities could not prohibit fracking, they have regulated it, oftentimes effectively, and found means through planning and local governance to reduce the externalities or improve local conditions. Sometimes policy makers and planners made mistakes. Other times, they have come up with workable and reasonable plans of action. They may not have a full range of choices, but they have some choice, and have often done the best with the lot they are given.

It is fitting to conclude with the call for more research, but to reiterate the importance of bringing together a diversity of perspectives to address the complex challenges arising with the shale boom and bust. The desire for expertise cannot come at the expense of working with practitioners. It is also critical to include at the bargaining table those folks who believe honestly that fracking provides needed economic opportunities. For example, just as someone might think it is a terrible idea to frack on a river basin, a farmer interested in leasing land to fracking companies may find the leasing option more appealing than selling the land to a developer who might turn the farm into a subdivision, or a Walmart. Taking seriously the idea of inclusive governance requires considering all voices, including those who support fracking. An inclusive process may be the best that can be hoped for when it comes to figuring out how to deal with the ups and inevitable downs of shale production.

References

Aldy, Joseph E., Scott Barrett, and Robert N. Stavins. 2003. "Thirteen plus One: A Comparison of Global Climate Policy Architectures." *Climate Policy* 3 (4): 373–97.

Allcott, Hunt, and Daniel Keniston. 2017. "Dutch Disease or Agglomeration? The Local Economic Effects of Natural Resource Booms in Modern America." *Review of Economic Studies* 85 (2): 695–731.

Boettke, Peter J., Emily Chamlee-Wright, Peter Gordon, Sanford Ikeda, Peter T. Leeson, and Russell Sobel. 2007. "The Political, Economic, and Social Aspects of Katrina." *Southern Economic Journal* 74 (2): 363–76.

Brennan, Jason. 2017. *Against Democracy*. Princeton: Princeton University Press.

Caplan, Bryan. 2006. *The Myth of the Rational Voter: Why Democracies Choose Bad Policies*. Princeton: Princeton University Press.

Chamlee-Wright, Emily, and Virgil Henry Storr. 2009a. "Club Goods and Post-Disaster Community Return." *Rationality and Society* 21 (4): 429–58.

Chamlee-Wright, Emily, and Virgil Henry Storr. 2009b. "'There's No Place like New Orleans': Sense of Place and Community Recovery in the Ninth Ward after Hurricane Katrina." *Journal of Urban Affairs* 31 (5): 615–34.

Cunningham, Scott, Gregory DeAngelo, and Brock Smith. Forthcoming. "Fracking and Risky Sexual Activity." *Journal of Health Economics*.

Gibbons, Ann. 2017. "Fracking Linked to Low-Weight Babies." *Science*, December 13. https://www.sciencemag.org/news/2017/12/fracking-linked-low-weight-babies#:~:text=They%20found%20that%20infants%20born,report%20today%20in%20Science%20Advances%20.

Granovetter, Mark S. 1977. "The Strength of Weak Ties: A Network Theory Revisited." *Social Networks* 1:201–33.

Koppl, Roger. 2018. *Expert Failure*. Cambridge: Cambridge University Press.

Mason, Charles F., Lucija A. Muehlenbachs, and Sheila M. Olmstead. 2015. "The Economics of Shale Gas Development." Discussion Paper 14-42-REV. Washington, DC: Resources for the Future. https://media.rff.org/documents/RFF-DP-14-42.pdf.

Mayfield, Erin N., Jared L. Cohon, Nicholas Z. Muller, Inês M. L. Azevedo, and Allen L. Robinson. 2019. "Cumulative Environmental and Employment Impacts of the Shale Gas Boom." *Nature Sustainability* 2 (12): 1122–31.

Ostrom, Elinor. 2010. "Polycentric Systems for Coping with Collective Action and Global Environmental Change." *Global Environmental Change* 20 (4): 550–57.

Piketty, Thomas. 2014. *Capital in the Twenty-First Century*. Cambridge, MA: Harvard University Press.

Putnam, Robert D. 2000. *Bowling Alone: The Collapse and Revival of American Community*. New York: Simon and Schuster.

Rickman, Dan S., Hongbo Wang, and John V. Winters. 2017. "Is Shale Development Drilling Holes in the Human Capital Pipeline?" *Energy Economics* 62:283–90.

Romer, Christina D. 1990. "The Great Crash and the Onset of the Great Depression." *Quarterly Journal of Economics* 105 (3): 597–624.

Romer, Christina D. 1992. "What Ended the Great Depression?" *Journal of Economic History* 52 (4): 757–84.

Saez, Emmanuel, and Gabriel Zucman. 2016. "Wealth Inequality in the United States since 1913: Evidence from Capitalized Income Tax Data." *Quarterly Journal of Economics* 131 (2): 519–78.

Sobel, Russell S., and Peter T. Leeson. 2006. "Government's Response to Hurricane Katrina: A Public Choice Analysis." *Public Choice* 127 (1–2): 55–73.

Somin, Ilya. 2020. *Free to Move: Foot Voting, Migration, and Political Freedom*. New York: Oxford University Press.

Storr, Virgil Henry, Laura E. Grube, and Stefanie Haeffele-Balch. 2017. "Polycentric Orders and Post-Disaster Recovery: A Case Study of One Orthodox Jewish Community Following Hurricane Sandy." *Journal of Institutional Economics* 13 (4): 875–97.

Troesken, Werner. 2015. *The Pox of Liberty: How the Constitution Left Americans Rich, Free, and Prone to Infection*. Chicago: University of Chicago Press.

Weimer, David L., and Aidan R. Vining. 2017. *Policy Analysis: Concepts and Practice*. 6th ed. New York: Routledge.

Contributors

Carla Chifos is associate professor, School of Planning, University of Cincinnati.

Teresa Córdova is director of the Great Cities Institute, and professor of urban planning and policy at the University of Illinois at Chicago.

Sabina E. Deitrick is associate professor of urban affairs and planning, Graduate School of Public and International Affairs, University of Pittsburgh.

Ann M. Eisenberg is associate professor of law, University of South Carolina School of Law.

Adelyn Hall is director of school-centered neighborhood development at the Community Learning Center Institute in Cincinnati.

Max Harleman is a recent PhD graduate of the Graduate School of Public and International Affairs, University of Pittsburgh.

Carolyn G. Loh is associate professor of urban studies and planning, Wayne State University.

Rebecca Matsco is board of supervisors chair, Potter Township (Beaver County, PA).

Larry McCarthy is professor of accounting, economics, and finance, School of Business, Slippery Rock University.

Nicholas G. McClure is a recent graduate of the Graduate School of Public and International Affairs, University of Pittsburgh.

Pamela A. Mischen is faculty adviser to the president, and associate professor, Department of Public Administration, Binghamton University.

Ilia Murtazashvili is associate professor in the Graduate School of Public and International Affairs, and associate director of the Center for Governance and Markets at the University of Pittsburgh.

Anna C. Osland is senior research associate, Kathleen Blanco Public Policy Center, University of Louisiana at Lafayette.

Erik R. Pages is president of EntreWorks Consulting, Arlington, VA.

Joseph T. Palka Jr. is professor in mathematics at Hudson Valley Community College, Troy, NY.

Ennio Piano is assistant professor of economics, Middle Tennessee State University.

Sandeep Kumar Rangaraju is assistant professor of economics, Weber State University.

Gavin Roberts is assistant professor of economics, Weber State University.

Heidi Gorovitz Robertson, J.D., J.S.D., is the Steven W. Percy Distinguished Professor of Law, Cleveland Marshall College of Law, and professor of environmental studies, Levin College of Urban Affairs, Cleveland State University.

Martin Romitti is senior vice president for research and technical assistance, Center for Regional Economic Competitiveness, Arlington, VA.

Ion G. Simonides is a recent graduate of the Graduate School of Public and International Affairs, University of Pittsburgh.

Frederick Tannery is associate professor emeritus in the School of Business, Slippery Rock University, and research associate at the University of Pittsburgh.

Jeremy G. Weber is associate professor, Graduate School of Public and International Affairs, University of Pittsburgh.

Mark C. White is associate extension professor and policy research scholar in the Truman School of Government and Public Affairs, University of Missouri.

Index

Page numbers in italics indicate tables and figures.